INTERNATIONAL SERIES OF MONOGRAPHS IN

ELECTRICAL ENGINEERING

VOLUME 6

Thyristor Control of A.C. Motors

OTHER TITLES IN THE SERIES IN ELECTRICAL ENGINEERING

Vol. 1. BEAN & SIMONS—*Lighting Fittings: Performance and Design*

Vol. 2. WILSON—*Modern Practice in Servo Design*

Vol. 3. KNIGHT—*Power Systems Engineering and Mathematics*

Vol. 4. JONES—*New Approaches to the Design and Economics of EHV Transmission Plant*

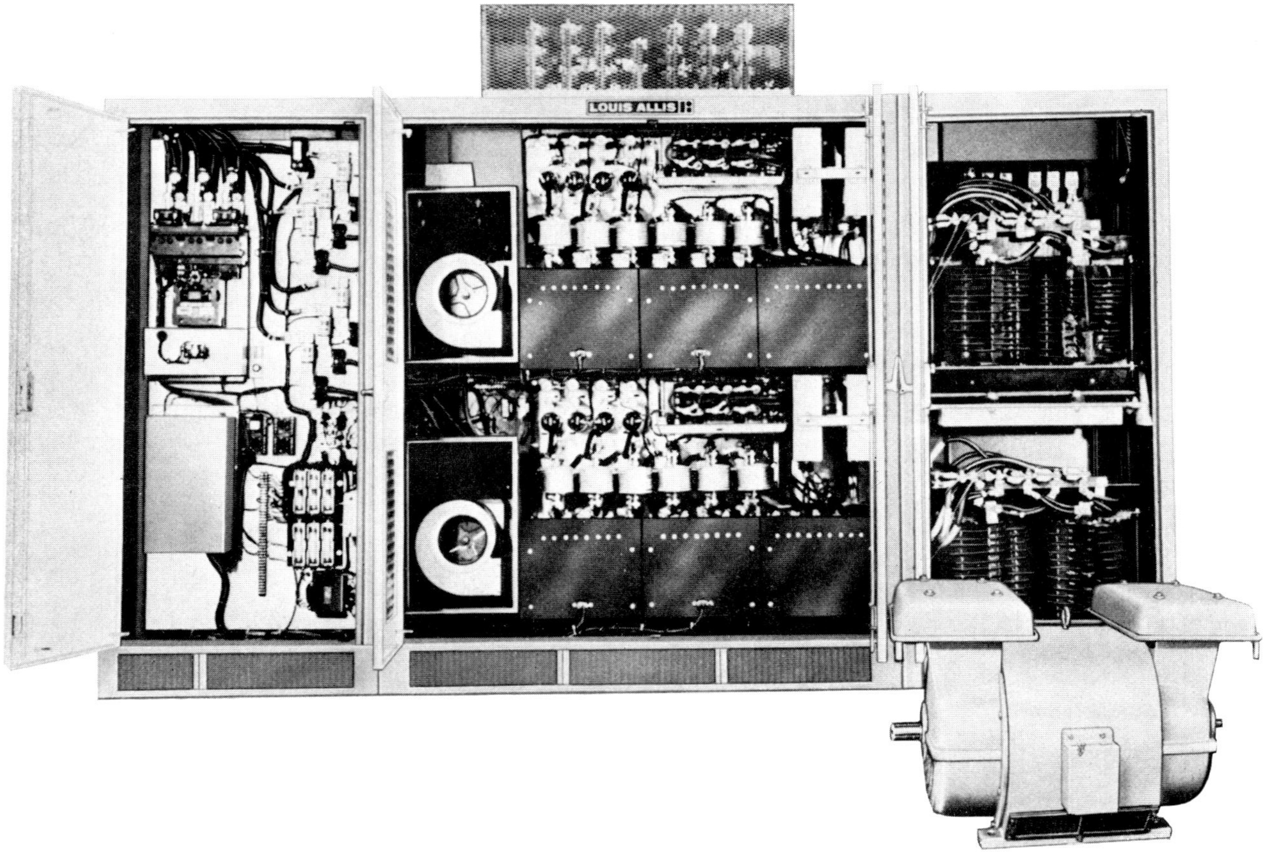

Frontispiece. A thyristor-controlled 500 horsepower variable-speed a.c. motor drive, in which a standard induction motor is powered by a variable-frequency current-source converter.
(Courtesy of Louis Allis Co., Milwaukee, U.S.A.)

Thyristor Control of A.C. Motors

by

J. M. D. MURPHY
M.E., Ph.D., C.Eng., M.I.E.E.
Department of Electrical Engineering, University College, Cork, Ireland

PERGAMON PRESS

OXFORD · NEW YORK · TORONTO
SYDNEY · BRAUNSCHWEIG

Pergamon Press Ltd., Headington Hill Hall, Oxford
Pergamon Press Inc., Maxwell House, Fairview Park, Elmsford, New York 10523
Pergamon of Canada Ltd., 207 Queen's Quay West, Toronto 1
Pergamon Press (Aust.) Pty. Ltd., 19a Boundary Street, Rushcutters Bay, N.S.W. 2011, Australia
Vieweg & Sohn GmbH, Burgplatz 1, Braunschweig

First edition 1973

Library of Congress Cataloging in Publication Data

Murphy, John M D
Thyristor control of A.C. motors.

(International series of monographs in electrical engineering, v. 6)
Includes biblioprahical references.
1. Electric controllers. 2. Thyristors.
3. Electric motors, Alternating current. I. Title.
TK2851.M8 1973 621.46′2 72–8043
ISBN 0-08-016943-0

Printed in Germany

ISBN 0 08 016943 0

Contents

Preface

THE thyristor or silicon controlled rectifier is a semiconductor device capable of controlling large currents. It was first introduced in 1957, and in the fifteen years since then it has revolutionized the field of electric power control and conversion. Power transistor ratings have also increased dramatically and, in recent years, the mercury arc rectifier, the motor-generator set and the saturable reactor have all been superseded by solid-state circuits using power diodes, power transistors and thyristors.

Solid-state power electronics has opened up new vistas in the field of motor control. Prototype battery-powered electric cars employ a thyristor-controlled electric motor. Thyristor drives have been widely used throughout industry and, for normal industrial applications, a d.c. motor powered by a thyristor converter is now a popular choice as a variable-speed drive. This text is concerned with solid-state variable-speed drives using a.c. rather than d.c. motors. The static variable-frequency a.c. drive uses a cage-rotor induction motor or synchronous reluctance motor powered by a static frequency converter. This gives a versatile and robust variable-speed machine which has the advantages over conventional variable-speed drives of higher accuracy, better reliability, reduced maintenance and higher efficiency. The main objection to the static a.c. drive has been on economic grounds, since the cost of the static frequency converter has often been considered excessive. However, power semiconductor prices are steadily decreasing as production volume grows and manufacturing techniques improve, and the future of the solid-state a.c. drive is assured.

The replacement of the d.c. machine by an a.c. motor also has economic benefits, since increasing labour and material costs are weakening the position of the d.c. motor with its elaborate commutator construction. The disadvantages of mechanical commutation are well known and the thyristor-controlled d.c. motor drive also produces a greater degree of a.c. mains distortion than certain types of solid-state a.c. motor drive.

Because of these factors, and a growing awareness of the performance possibilities, a more widespread application of the solid-state a.c. drive is inevitable. When extremely precise speed control or precise speed matching is demanded, the solid-state a.c. drive is being used to provide standards of accuracy and reliability which have never before been achieved, and it is now an attractive proposition at power levels from a few kilowatts to several megawatts. The improving economic position of the solid-state a.c. drive also means that, in the future, it will be increasingly considered for general purpose applications.

This book is intended as an introduction to the basic theory and practice of static power conversion using thyristors and, in particular, it deals with the application of such circuits for the speed control of a.c. motors. The text is suitable for degree-level and post-graduate courses in Power Electronics and Electrical Machines. It is also intended as an introductory text for practising engineers who wish to familiarize themselves with recent developments in the rapidly expanding field of thyristor control and who may need to specify, purchase or use solid-state a.c. drives. It is hoped that the text will also serve as a useful reference work for engineers already active in the field.

In Chapter 1, the fundamental principles of a.c. motor operation are introduced and in Chapter 2 basic rectifier and inverter theory is developed. In addition, two appendices are included: Appendix A on fundamental induction motor theory and Appendix B on basic thyristor theory. In this manner, it is intended as far as possible to make the book self-contained, and it is expected that the text will prove equally intelligible to an engineer or student with a background in either electrical machines or electronics.

A glance at the table of contents will indicate the scope of the text. It includes such topics as: static inverter and cycloconverter theory; the effect of supply harmonics on a.c. motor performance; stability of converter-fed a.c. motor drives; controlled-slip operation of the induction motor; the effect of converter loading on the a.c. supply; the static inverter specification procedure, etc. When a somewhat lower standard of performance is adequate, a satisfactory but less expensive solid-state a.c. drive can be obtained by means of stator voltage control of the induction motor or slip-energy recovery with a wound-rotor induction motor. These techniques are also treated in the text.

Particular aspects of the subject-matter have been treated in various journals and conference proceedings but this book is an attempt to produce a comprehensive account of theory and practice in the field. To the author's knowledge, no such textbook has been published. An extensive list of references has been included for those who may require further information on the various aspects of the subject.

I gratefully acknowledge the help of those who have encouraged and assisted me in the preparation of this book. I am particularly indebted to Professor C.T.G.Dillon, of University College, Cork, who, in the year 1960 introduced me to the thyristor. I also wish to thank Professor J.C.West of Sussex University, who first suggested that I write a book on thyristor control of a.c. motors. I am grateful to Miss Mella Kilbride for her excellent typing of the manuscript, and last but not least, I wish to record my gratitude to my wife, Margaret, for her patience during the many hours required for the preparation of the text.

University College, Cork JOHN M.D.MURPHY

CHAPTER 1

Principles of Variable-frequency A.C. Drives

1.1. INTRODUCTION

In many modern variable-speed drives the demand is for a precise and continuous control of speed with long-term stability and good transient performance. The d.c. motor has satisfied these requirements satisfactorily in many applications, but the mechanical commutator is often undesirable because of the regular maintenance required. This causes difficulty when service interruptions cannot be tolerated or when the motor is used at inaccessible locations. A.C. motors such as the squirrel-cage induction motor and the synchronous reluctance motor have a robust rotor construction which permits reliable maintenance-free operation at high speed. The simple rotor construction also results in a cheaper motor and a higher power/weight ratio. Unfortunately, the induction motor and reluctance motor are both inflexible in speed when operated on a standard constant-frequency a.c. supply. The reluctance motor operates synchronously at a speed which is determined by the supply frequency and the number of poles for which the stator is wound. The induction motor runs slightly below synchronous speed. For intermittent operation at reduced speeds, stator voltage control of the induction motor is satisfactory (Chapter 9). Sub-synchronous speed control of a wound-rotor induction motor is obtained by means of a converter cascade for slip-energy recovery (Chapter 10). However, efficient wide-range speed control of the reluctance motor or cage-rotor induction motor is only possible when a variable-frequency a.c. supply is available. Consequently, in this text, attention is largely concentrated on the variable-frequency method of obtaining variable-speed a.c. motor operation.

Hitherto, the various techniques for the speed control of a.c. motors often required auxiliary rotating machines. Nowadays, static a.c. drives, using thyristors (silicon controlled rectifiers) as solid-state switching devices, have been developed and marketed in Europe and in the United States of America.[1,2] Falling thyristor prices and an appreciation of the performance possibilities have created a growing interest in solid-state a.c. drives, and consequently the variable-frequency drive, in particular, is being increasingly applied in industry.

In a static frequency converter, the fixed-frequency a.c. supply voltage is transformed to a variable-frequency output which is used to power conventional a.c. motors. In this text, attention is concentrated on static frequency converters using thyristors, but much of the text is also applicable to transistorized converters. Power transistor ratings of tens of amperes and several hundred volts are now available and, for an inverter rating of less than 5 kVA, a transistorized inverter may be cheaper than a thyristor inverter. Consequently, transistor inverters are economically competitive for a.c. motor drives of less than 5 h.p., but thyristors are essential for higher powers.

1.2. SELECTING A VARIABLE-SPEED DRIVE

The problem in selecting a variable-speed drive for a particular application is to choose the system which can most economically provide the required range of speed control with the desired accuracy and speed of response. The a.c. commutator motor has been widely used, since it can be supplied directly from the a.c. mains, but for a reversible drive with continuous speed control over a very wide range the d.c. motor has been the most popular solution. The separately excited d.c. motor is rapidly and efficiently controlled by variation of the armature voltage and field current. In recent years the d.c. supply has been obtained from the a.c. network by means of static converters which permit the controlled rectification of the alternating voltage so that a variable direct voltage is provided for the armature. Precise speed control is achieved by adopting closed-loop feedback methods.

However, the d.c. motor is not the ideal solution to the problem of variable-speed motor operation. The commutator consists of a large number of copper segments separated by thin sheets of mica insulation. This elaborate construction increases the cost of the d.c. motor and reduces the power/weight ratio. Brush and commutator wear are accentuated by sparking, and the mica insulation limits the voltage between segments. The total armature voltage is therefore limited to a maximum of about 1500 V. The magnitude of the armature current and its rate of change are restricted by commutation difficulties, and the speed of rotation of the d.c. machine is limited. The squirrel-cage induction motor, on the other hand, has a rotor circuit consisting of a short-circuited winding which can often be made from a single casting. There is no necessity to insulate the rotor bars from the surrounding laminations, and the cage rotor has a low inertia and can operate at high temperature and high speed for prolonged periods without maintenance. In addition, the cost of the cage-rotor induction motor is only about one-sixth of that for a d.c. motor of the same speed and horsepower rating. The power/weight ratio of the squirrel-cage motor is about twice that of the d.c. machine, and induction motors are manufactured in much larger power ratings since the stator current is not limited by commutation, and the stator voltage can be 15 kV or more. It is obvious, therefore, why numerous attempts have been made to obtain economic and efficient speed control of the cage-rotor induction motor. A certain degree of speed variation can be achieved by reduction of the stator voltage at constant supply frequency, but for efficient wide-range speed control the stator frequency must be varied.

The provision of a variable-frequency supply by rotating-machine sets or static frequency converters is expensive. Consequently, the high cost of the frequency conversion equipment often offsets the saving in cost due to the replacement of the d.c. motor by the cheaper cage-rotor machine. However, the static frequency converter provides a variable-speed drive whose performance in terms of accuracy and reliability has never before been achieved, and the initial capital cost is frequently justified by the improved performance and lower running costs. The variable-frequency drive is particularly attractive in multi-motor systems when large numbers of small motors are supplied simultaneously with the same frequency. In such applications, the cost of the frequency converter is justified by the significant reduction in machine costs due to the large number of motors involved. If these motors are required to run at closely co-ordinated speeds, the variable-frequency system is highly advantageous since modern static frequency converters can deliver an output frequency of extremely high precision and long-term stability. When this precise frequency output is used to power synchronous or reluctance motors, the accuracy and stability are reproduced in the shaft speed. Exact speed ratios may be obtained between individual motors by using different gear ratios or pole numbers. Alternatively, digital control techniques may be adopted.[3] Permanent-magnet synchronous motors or reluctance motors are used in order to retain the advantages of a brushless construction. Modern static drives of this type can readily provide speed holding accuracies of 0·001 per cent, or better, at any set speed and are superior to any other method available. Such multi-motor drives have found widespread application in the textile and synthetic fibre industries, in the paper-making industry and in various processing lines where exact speed co-ordination is essential in order to maintain the quality of the product.[4] In these applications, reliable operation is also vital since a system failure can lead to substantial production losses. This reliability is ensured by a static frequency converter and standard brushless a.c. motors. Single-motor variable-frequency drives have been used where alternative speed control methods do not provide sufficient accuracy of control. They are also used in adverse industrial environments and in applications where a commutator motor is unacceptable because of its maintenance requirements.

The static frequency converter includes much specialized electronic circuitry and its maintenance might appear to present some difficulty. However, by adopting techniques of modular construction with removable units, the faulty circuit unit can be readily located and replaced. The repair of the faulty module can then be carried out by the skilled maintenance staff under more favourable conditions. The reliability of the electronic control circuitry can be improved by adopting monolithic integrated circuit techniques. The thyristors themselves have unlimited life provided the manufacturer's specifications are not exceeded. The transient current rating is limited and must be carefully observed. This may require oversizing of the thyristor converter so that it can deliver the high starting current of the motor. This aspect is discussed more fully in Chapter 7.

When first developed, the cost of thyristor devices was prohibitive but it is now falling steadily as production grows. A.C. speed control systems using static frequency converters are, therefore, becoming increasingly competitive with more established methods.[5] The proof of this competitiveness lies in the widespread application of such equipment and in the wide range of units, from a few kVA to several hundred kVA, which is now available from various manufacturers. In future years, as circuit development continues and thyristor costs fall, there is no doubt that the use of static frequency converters for variable-speed applications will continue to grow in importance.

1.3. CONSTANT VOLTS/HZ OPERATION

A static converter which delivers variable-frequency power to a motor or transformer must also vary the terminal voltage as a function of frequency in order to maintain the proper magnetic conditions in the core. In practice, magnetic devices usually operate near saturation in order to give maximum utilization of the core material. When the operating frequency is reduced, the applied voltage must be reduced proportionately or the saturation flux density is exceeded, resulting in excessive iron losses and magnetizing current. When the operating frequency is increased, the applied voltage should be increased proportionately in order to maintain the magnetic flux density. This may be understood when it is realized that the e.m.f. induced in a winding is proportional to the rate of change of magnetic flux. As the operating frequency increases, the rate of change of flux also increases, and for a constant flux amplitude the induced e.m.f. increases linearly with frequency. When an alternating voltage is applied to a motor or transformer winding of negligible resistance, a counter e.m.f. must be induced which is equal and opposite to the applied voltage at every instant. The pulsating flux which links the winding automatically adjusts its amplitude so that this condition is satisfied. This means that, in order to maintain constant flux, the applied voltage and counter e.m.f. should increase linearly with frequency or the applied voltage/frequency ratio must be held constant. This mode of operation is known as constant volts/Hz or simply constant "volts/cycle".

Many variable-speed drives require a constant-torque output and this can be achieved if the air-gap flux in the motor is maintained constant by operating on a fixed volts/Hz supply. However, the analysis above assumes negligible winding resistance, whereas, in practice, at low frequencies the resistive potential difference becomes significant compared with the induced e.m.f. This causes a reduction in the air-gap flux and motor torque. In order to maintain the low-speed torque, the volts/Hz must be increased at frequencies below about 20 Hz.

1.4. GENERATION OF VARIABLE-FREQUENCY A.C. POWER

The idea of using a variable-frequency supply to control the speed of a.c. motors is not new, and rotating frequency converters have been employed for many years. These are used principally in multi-motor mill drives and in special applications where a high operating frequency is chosen in order to permit the use of compact a.c. motors. Nowadays the rotating machine methods of variable-frequency a.c. power generation are being largely supplanted by static conversion methods.

1.4.1. *Rotating Frequency Converters*

The rotating frequency converter may consist of a synchronous alternator driven at variable speed in order to generate variable-frequency power. If the field current is held constant, the alternator delivers a constant volts/Hz output since both output voltage and frequency are proportional to speed. By increasing the field current at low speeds, the volts/Hz can be boosted for low-frequency operation. The alternator is driven at variable speed by a d.c. motor or a.c. commutator motor and the rotating-machine set is sited in a convenient position where commutator maintenance can be carried out. The speed-controlled a.c. motors are placed in more hazardous and inaccessible locations.

1.4.2. *Static Frequency Converters*

The performance and reliability of a variable-speed a.c. system are improved if the rotating frequency converter is replaced by static methods of variable-frequency power generation.[6-13] In order to obtain high efficiency, it is essential to use static switching devices which are either "on" or "off". In the on-condition, the device approximates to an ideal closed switch having zero voltage drop across it and a current which is determined by the external circuit. In the off-condition, the device approximates to an ideal open switch which has infinite impedance and blocks the flow of current in the circuit. If the solid-state switch can be triggered from the off-condition into the on-condition by a low-power control signal, the device can be used in converter circuits for the generation of variable-frequency alternating voltages. The grid-controlled mercury arc rectifier and the thyratron can be used in this manner, and many of the circuits used nowadays were originally developed using these devices. However, these early circuits were not sufficiently attractive technically or economically and were not adopted commercially. The development of the thyristor, which is the semiconductor counterpart of the thyratron, and its availability in high-power ratings, have given a new boost to static methods of variable-frequency a.c. generation and many of the earlier circuits have been revived and improved.

As in the case of the thyratron, the thyristor can be triggered into conduction by an external signal on the trigger or gate electrode. However, like the thyratron, the conducting thyristor cannot be switched off by an external signal. Turn-off or commutation, as it is called, can only be achieved by interrupting the anode current of the thyratron or thyristor. If a thyristor circuit operates on an a.c. supply, the natural reversal of the alternating voltage every half-cycle may be used to reduce the thyristor current to zero and so produce natural commutation. In a d.c.-operated circuit the current does not go to zero naturally, and forced commutation must be used.[14,15] This technique usually employs auxiliary charged capacitors to force the current in a conducting thyristor to zero.

The renewed interest in static frequency conversion is due to the improved characteristics of the thyristor as compared with the thyratron. The thyristor is a more efficient switching device since the voltage drop in the on-condition is only about 1 V as compared with 20 V, or more, for the thyratron and mercury arc valves. In addition, the switching times of the thyristor are an order of magnitude lower than those of the thyratron. The reduced turn-off time permits a significant reduction in the auxiliary apparatus necessary to achieve forced commutation. The thyristor is also a more rugged and durable device and is more compact, even with its associated heat sink. Being a semiconductor device, the thyristor has no warm-up time and it eliminates the filament heating supply of the thyratron.

There are two basic types of static frequency converter, the d.c. link converter and the cycloconverter. A block diagram of the d.c. link converter is shown in Fig. 1.1. The three-phase a.c. supply is first rectified to d.c. in a standard rectifier and the resulting d.c. power is fed to a static inverter. An inverter is a device which converts d.c. power to a.c. power, and the static inverter uses thyristors (or transistors) which are switched sequentially so that an alternating voltage waveform is delivered to the a.c. motor. Many suitable inverter circuits have been developed and are described in the literature.[16–38] The output frequency is determined by the rate at which the inverter thyristors are triggered into conduction. This is controlled by the reference oscillator and logic circuits, which generate and distribute firing pulses in the correct sequence to the various thyristors. At the end of its conduction period, each thyristor must be turned off by an auxiliary commutating circuit. Static inverter circuits are considered in more detail in Chapters 2 and 3.

The inverter output frequency may be controlled from zero to several hundred Hz by variation of the reference oscillator frequency. The output voltage waveform is usually non-sinusoidal but external filter circuits are not employed due to the difficulty of obtaining effective operation over a wide range of frequencies. The inverter output is therefore supplied

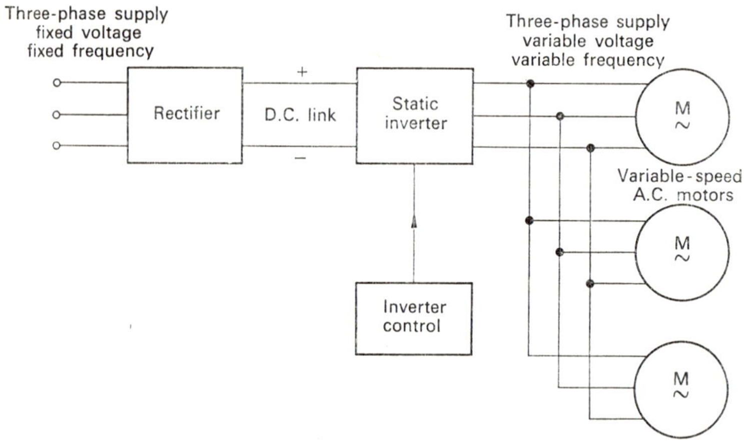

FIG. 1.1. Block diagram of a d.c. link converter for a multi-motor a.c. drive.

directly to the a.c. motor, and the harmonic effects must be taken into consideration when specifying the motor (Chapter 7). Normally it is found that the distorted waveform does not impose any serious limitations apart from a slight reduction in the rating and efficiency of the motor. The inverter-operated a.c. motor also exhibits a non-uniform stepping motion at very low speeds due to the non-uniform rotation of the flux wave. The volts/Hz output of the inverter can be controlled by variation of the direct voltage input to the inverter. Alternatively, the input voltage may be held constant while the volts/Hz ratio is adjusted within the inverter circuit. Various voltage control techniques are available and are considered in Chapter 3.

The second basic form of static frequency converter is the cycloconverter and this has also been used in variable-speed a.c. drives.[39–45] In a cycloconverter, the network frequency is converted directly to a lower output frequency without intermediate rectification. Thyristors are used to selectively connect the load to the supply source, and the low-frequency output voltage waveform is fabricated from segments of the supply voltage waveform. The output frequency must be less than about one-third of the input frequency, and the drive is therefore only suitable for low-speed motor operation if the input is at normal mains frequency. However, the output voltage waveform approximates closely to a sine wave, particularly at low frequencies. As in the static inverter, the output frequency is determined by an independent reference oscillator, and the output volts/Hz is also varied by the control circuitry. The cycloconverter is treated more fully in Chapter 4.

A three-phase a.c. motor is reversed by changing the phase sequence of the supply voltage. In a static converter this is achieved by reversing the firing sequence of thyristors in the static inverter or cycloconverter circuit, and no switching of power leads is necessary. One of the principal advantages of the cycloconverter is that it is inherently capable of regenerative operation. This means that the direction of energy flow can be reversed so that energy is returned to the supply from the low-frequency side. This permits rapid motor deceleration by returning the kinetic energy of the rotating parts to the a.c. supply. Regenerative braking is more satisfactory than dynamic braking in which the kinetic energy is dissipated in resistance losses. The regenerative method is more readily controlled and provides operating economy.

The d.c. link converter involves a double power conversion, but efficiencies of 85 to 95 per cent are achieved in practice, and the output frequency is chosen to suit the application. However, the basic d.c. link converter cannot operate regeneratively unless the d.c. link can return energy to the a.c. network. This is not possible with a simple rectifier connection and additional circuitry is required which adds to the cost and complexity of the system. In many applications, regenerative braking is not necessary and the d.c. link converter then provides a variable-speed drive which is highly competitive with more conventional methods.

1.4.3. *Advantages of Static Frequency Conversion*

Static frequency converters have many advantages over rotating frequency converters, although the initial cost of the static system is usually greater, particularly when regenerative braking is required. However, the lower maintenance requirements, the reduction in downtime and the improved product quality often justify the higher capital cost.

In a rotating frequency converter using a motor-alternator set, the application of load to the controlled motors imposes an increased load on the alternator and its driving motor, thereby reducing output voltage and frequency. Similarly, when several motors are operating on the same variable-frequency supply, the starting-up of an additional motor will cause a momentary drop in the voltage and frequency delivered to the other motors. In order to obtain precise speed control with a rotating frequency converter, closed-loop tacho-generator feedback must be used. High accuracy demands an over-sized alternator and a high loop gain, with the attendant risk of system instability. In a static frequency converter, as already explained, the output frequency is determined solely by the independent reference oscillator. The output frequency is therefore completely independent of fluctuations in the a.c. supply frequency and voltage and is also independent of load variations. Consequently, zero frequency regulation is obtained. When the static frequency converter supplies an induction motor, the shaft speed departs from synchronous speed by an amount dependent on the load torque, but when a synchronous motor or reluctance motor is used in a simple open-loop system, the motor speed is determined solely by the reference oscillator. Extremely high accuracy and stability are obtained, and the speed regulation on load is zero, although sudden changes in load can produce momentary deviations from synchronous speed.

Other advantages of static frequency conversion are:

1. Static converters have lower installation costs since they do not require elaborate foundations or careful machine alignment. Static equipment also has smaller space requirements and a lower noise level.
2. The operating costs of a static converter are lower due to the higher efficiency and the absence of moving parts which deteriorate with time and require periodic replacement.

3. The static converter has greater ease of control since the output voltage and frequency can be independently varied over a wide range, and closed-loop feedback methods are readily applied. The output volts/Hz can be adjusted to suit the variable-speed motors and, if necessary, very large starting torque can be developed.

1.5. VARIABLE-FREQUENCY OPERATION OF A.C. MOTORS

The static frequency converter usually supplies a squirrel-cage induction motor or a reluctance motor. Conventional a.c. motor operation at fixed frequency is now briefly reviewed and compared with variable-frequency operation. The basic induction motor formulae, which will be used in the text, are derived in Appendix A, and variable-frequency operation is treated more fully in Chapter 6.

1.5.1. *The Squirrel-cage Induction Motor*[(46)]

In the induction motor, the rotor winding is short-circuited and receives its supply by induction from the stator. The stator carries a three-phase winding which is directly connected to a three-phase a.c. supply. This establishes an air-gap magnetic field of constant amplitude which rotates at synchronous speed $n_1 = 60f_1/p$ rev/min (p is the number of pole-pairs for which the stator is wound and f_1 is the stator frequency in c/s or Hz). The synchronous speed of the motor can, therefore, be controlled by variation of the supply frequency, f_1. The amplitude of the air-gap flux is determined by the stator volts/Hz as already explained in Section 1.3. When the induction motor is supplied at fixed voltage and frequency, the air-gap flux is approximately constant for all normal loading conditions. The phase sequence of the stator supply determines the direction of rotation of the air-gap field and hence determines the direction of rotation of the motor. The motor runs at a speed, n, which is usually only a few per cent less than the synchronous speed, n_1. As a result, the rotor e.m.f. is at a frequency $f_2 = sf_1$, where the fractional slip, $s = f_2/f_1 = (n_1 - n)/n_1$. The rotor current, I_2, lags the e.m.f. by the rotor phase angle, ϕ_2, and, as shown in Appendix A, the motor torque is proportional to the in-phase rotor current, $I_2 \cos \phi_2$. The output torque is also proportional to the air-gap flux Φ, and, in general

$$T = K\Phi I_2 \cos \phi_2, \tag{1.1}$$

where K is a constant of proportionality.

Figure 1.2 shows the torque-speed characteristic of a typical cage-rotor induction motor operating on a standard fixed-frequency supply. At synchronous speed the motor develops zero torque, and the speed decreases linearly with torque when the slip is small. As the rotor slip frequency increases, the rotor leakage reactance becomes significant and increases the rotor impedance and also the phase angle ϕ_2. With increasing slip, the motor torque therefore reaches a maximum and then diminishes. The maximum torque is called the breakdown torque, since the motor stalls if the load torque exceeds this peak value. The rotor frequency at the breakdown point is called the rotor breakdown frequency.

If the motor rotates backwards against the forward-rotating field the slip is greater than unity, and a counter-torque is developed which opposes the rotation. The motor operates in this region of the characteristic if two stator leads are suddenly reversed when the machine is operating normally as a motor. This changes the phase sequence of the stator currents and hence reverses the direction of rotation of the air-gap field. The machine is brought rapidly to rest and, if the supply is not disconnected, the rotor then speeds up in the opposite direction. This method of braking or rapidly reversing the induction motor is known as plugging.

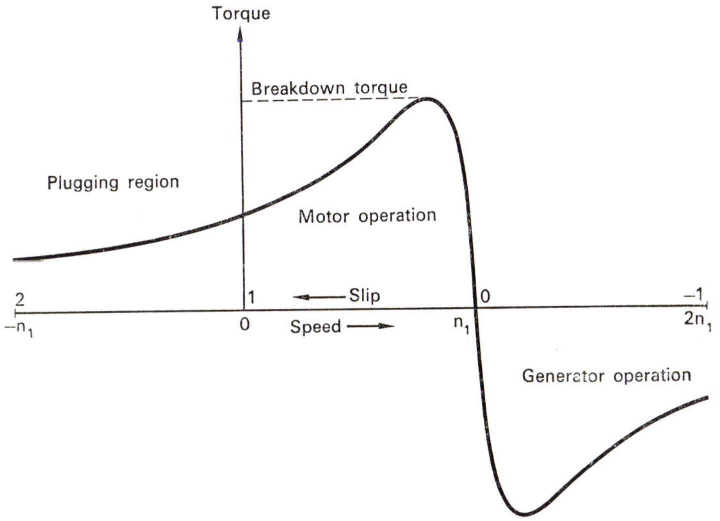

FIG. 1.2. Torque-speed characteristic of the polyphase induction motor at constant voltage and frequency.

The torque-speed curve may also be extended into the generator region above synchronous speed. In this region, the slip is negative and the machine converts mechanical energy at the shaft into electrical energy which is returned to the a.c. network. The machine operates under these conditions when the rotor is driven mechanically at super-synchronous speeds. Generator operation also occurs in a variable-frequency system when the supply frequency is rapidly reduced. The motor is then overhauled by the load, and the kinetic energy of the rotating masses is converted into electrical energy, giving regenerative braking. If the motor is supplied by a static frequency converter, the regenerated power must be returned to the supply through the converter.

The standstill current of a typical squirrel-cage induction motor is 5 or 6 times rated current, but the starting torque is small because of the low rotor power factor at high rotor frequencies. In a variable-frequency system, the supply frequency is reduced for starting. This improves the rotor power factor and so increases the torque per ampere at starting. The air-gap flux can also be increased by adjusting the stator volts/Hz. In this manner, several times rated torque is available at standstill, and the induction motor is rapidly run-up to speed by increasing the supply frequency. With this method of starting, there is no danger of the low-frequency crawling which sometimes occurs when induction motors are started on a fixed-frequency supply.

Figure 1.3 shows the torque characteristics of the motor at several different stator frequencies. The breakdown torque is maintained constant by adjusting the stator volts/Hz so that the air-gap flux is constant. These motor characteristics are suitable for driving a constant-torque load at variable speed. The difference between the synchronous speed, n_1, and the shaft speed, n, is the slip speed in rev/min and is practically constant over the speed range for a given load torque. Thus, if the motor has a 40 rev/min slip at rated load and 50 Hz, it will also have approximately 40 rev/min slip at 25 Hz. The shaft speed is, therefore, not quite proportional to the supply frequency but is near enough for many applications.

If the stator voltage remains constant as the frequency is varied, the air-gap flux and breakdown torque decrease with increasing frequency as shown in Fig. 1.4. These characteristics are suitable for traction applications where a large torque is required at standstill and low speeds, and a smaller torque is sufficient for high-speed running.

The shape of the induction motor torque-speed characteristic is largely determined by the rotor resistance and, for normal operation at a fixed frequency, the value of rotor resistance is a compromise. The rotor losses are proportional to slip, and a low-resistance rotor has a small slip and high efficiency at normal loads but it also permits excessive starting current

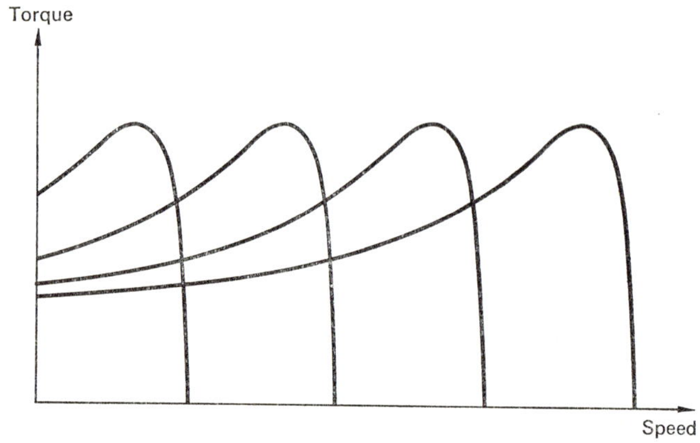

FIG. 1.3. Induction motor torque-speed characteristics at different supply frequencies and constant air-gap flux.

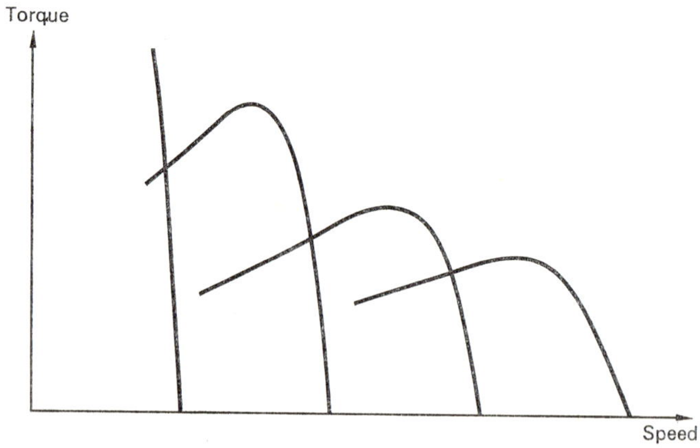

FIG. 1.4. Induction motor torque-speed characteristics at different supply frequencies and constant terminal voltage.

and yields a low starting torque. A high-resistance rotor, on the other hand, improves the starting conditions, but the full-load slip is large and operating efficiency is reduced. When the induction motor operates on a variable-frequency supply, the stator frequency can be reduced for starting, and the normal starting conditions at fixed frequency are not applicable. A low value of rotor resistance is therefore chosen in order to improve the operating efficiency. A low-resistance rotor is also desirable in order to reduce the harmonic losses on non-sinusoidal supplies, since the rotor copper loss is one of the main sources of additional harmonic losses (Chapter 5).

In general, induction motor operation takes place at high power factor and high efficiency provided the rotor breakdown frequency is not exceeded. Beyond the breakdown point, the motor currents are large but the power factor is low and the torque per ampere and efficiency are poor. The induction motor can be operated in a closed-loop feedback system in which the rotor slip frequency is controlled, so that operation always takes place below the breakdown frequency. This type of induction motor drive has speed control and transient performance characteristics which are superior to those of a d.c. motor supplied by a Ward–Leonard set or a static converter. The controlled-slip mode of operation is treated in Chapter 6.

1.5.2. *The Synchronous Reluctance Motor*

The elementary synchronous reluctance motor has an unexcited ferromagnetic rotor with polar projections. The reluctance motor, therefore, differs from the conventional synchronous machine which has a d.c.-excited winding on the rotor. In general, reluctance torque is developed by the tendency of a ferromagnetic material to align itself with a magnetic field. Figure 1.5 shows a two-pole rotor construction which could be used in a reluctance motor.

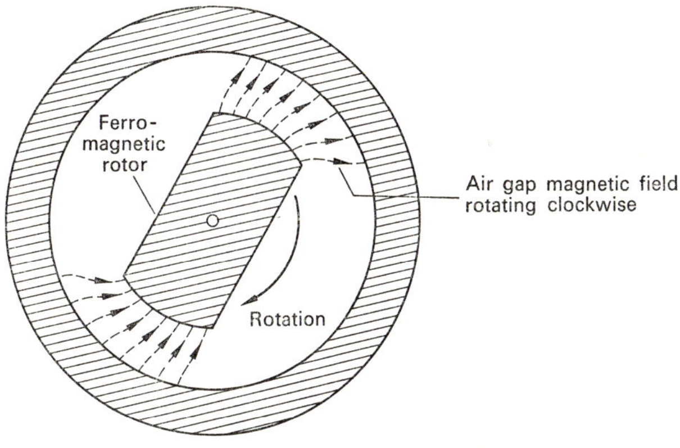

FIG. 1.5. Elementary two-pole reluctance motor.

If the stator establishes a stationary magnetic field, the salient-pole rotor experiences a torque tending to rotate the polar or direct axis into alignment with the stator field axis. This is the position of minimum reluctance for the stator flux. In practice, the stator carries a conventional three-phase winding excited by a three-phase a.c. supply. A synchronously rotating stator field is established, and the rotor runs in exact synchronism with this field as the salient poles try to maintain the minimum reluctance position relative to the stator flux. The rotor is also fitted with a squirrel-cage winding and the machine runs up from standstill by induction motor action.

As the rotor speed approaches synchronism, the reluctance torque is superimposed on the induction motor torque and, as a result, the rotor speed oscillates about its average value. Provided the load torque and inertia are not excessive, the instantaneous rotor speed can increase to synchronous speed and the rotor then locks into synchronism with the stator field. Figure 1.6 shows a typical torque-speed characteristic at a fixed supply frequency. At sub-

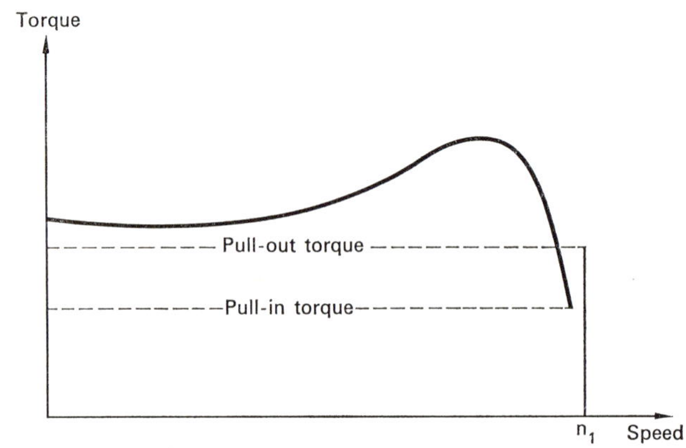

FIG. 1.6. Torque-speed characteristic of the reluctance motor at constant voltage and frequency.

synchronous speeds the torque is pulsating and the average value is indicated in Fig. 1.6. Once the rotor is synchronized, the cage winding rotates synchronously with the stator field and so the rotor winding plays no part in the steady-state synchronous operation of the motor. The machine continues to operate synchronously provided the pull-out torque of the motor is not exceeded. This is the load torque required to pull the rotor out of synchronism. The pull-in torque is defined as the maximum load torque which the rotor can pull into synchronism with a specified load inertia. The pull-in torque may be increased at the expense of a larger starting current but it is always less than the pull-out torque.

Up to recent years, the reluctance principle had only been applied to fractional horsepower motors, since the advantages of a reliable and robust construction seemed to be offset by the low efficiency, low power factor, poor torque/weight ratio and high starting current. However, recent development effort has succeeded in producing better reluctance motors.(47–50) Motor ratings of 150 h.p. with improved performance characteristics have been achieved by adopting a segmented-rotor construction.(51,52) In general, the synchronous performance is improved by increasing the degree of magnetic asymmetry in the rotor. Thus, a reduction in the flux on the interpolar or quadrature axis gives a greater pull-out torque, but the pull-in torque of the motor is reduced and the power factor is also affected. For fixed-frequency operation, the final design is a compromise between these factors. Modern reluctance motors have been constructed which have a power output of 75 to 85 per cent of that of an induction motor of the same size. Power factors of 0·65 to 0·75 have been reported with a full-load efficiency of 85 per cent.[53] An improvement in the pull-in performance is obtained at the expense of the power factor and efficiency, and segmented-rotor reluctance motors are now available which can synchronize up to 10 times the rotor inertia against full-load torque, with a starting current of only 4·5 times the full-load current. The pull-out torque is typically about 1·5 times the full-load torque.

When the supply voltage and frequency are varied, a family of torque-speed curves is obtained as in the case of the induction motor. For constant-torque applications, the volts/Hz must again be held approximately constant with a voltage boost below 20 Hz. Reluctance motors have been widely used in variable-speed multi-motor drives requiring exact speed co-ordination between individual motors. If all motors are accelerated simultaneously from standstill by increasing the supply frequency, the machines operate synchronously at all times and they can be designed for optimum synchronous performance without regard to the pull-in torque requirements. Unfortunately, the reluctance motor also exhibits a tendency towards instability at low supply frequencies (Chapter 5), but it forms a cheap, robust and reliable synchronous machine and its further development is being stimulated by the improved performance of the modern static frequency converter.

CHAPTER 2

Introduction to Static Rectification and Inversion

2.1. THE THYRISTOR[54–57]

Static power conversion can be achieved by means of controlled rectifiers such as the thyratron, the grid-controlled mercury arc valve, or the modern semiconductor equivalent of these devices, the thyristor.

The thyristor or silicon controlled rectifier was first introduced in 1957 by the General Electric Company of America. It is a static switching device which may be triggered from a blocking condition into a conducting state by means of a low-power signal at the control electrode. In the blocking condition, the thyristor corresponds to an open switch and current flow is negligible. When it is fully conducting, it behaves like a closed switch and the current flow is limited only by the external circuit. This simple picture of thyristor behaviour does not explain the functions of the auxiliary gating and commutating circuitry and in order to understand these aspects we must examine the physical behaviour of the device.

An ordinary silicon rectifier diode consists of a *p–n* junction of which the *p*-layer is the

FIG. 2.1. Thyristor appearance, structure and graphical symbol.

anode. The thyristor, as shown in Fig. 2.1, is a four-layer arrangement of *p–n–p–n* silicon. The end *p*-layer forms the anode, and the end *n*-layer is the cathode, while the *p*-layer adjacent to the cathode forms the gate or trigger electrode. The reverse characteristic is similar to that of the silicon diode, as it draws only a small leakage current when negative anode-to-cathode voltage is applied. If the critical reverse breakdown voltage is exceeded, a destructive breakdown occurs.

When positive anode-to-cathode voltage is applied, the thyristor again draws a small leakage current until the critical forward breakover voltage, $V_{(BO)}$, is exceeded. The thyristor then switches rapidly into the fully conducting condition. However, this triggering mechanism should not be used unless permitted by the manufacturer.

2.1.1. *Gate Triggering*

The thyristor is normally triggered by injecting a small external current which forward-biases the gate-cathode junction, J_3. This reduces the forward breakover voltage as shown in Fig. 2.2. Consequently, the thyristor may be triggered by holding the external forward

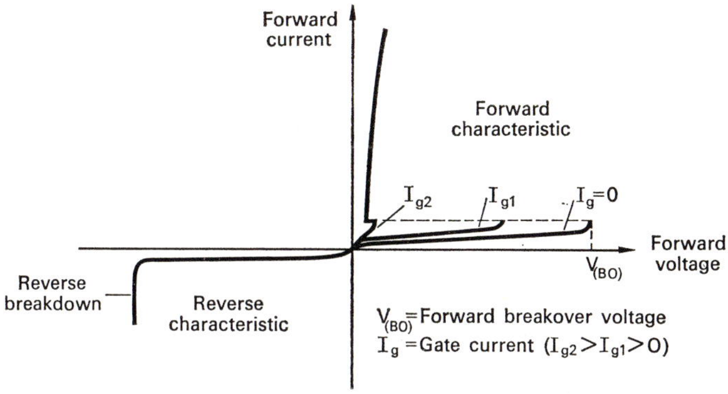

FIG. 2.2. Anode-to-cathode characteristics of the thyristor.

voltage below the breakover value and then, at the desired instant, injecting sufficient gate current to reduce the breakover voltage below the applied voltage. It is shown in Appendix B that the thyristor can be visualized as consisting of two interconnected complementary transistors, a *p–n–p* and an *n–p–n*. Using the two-transistor analogy, the mechanism of thyristor firing by gate triggering or forward breakover can be explained.

Once the thyristor conducts, the gate electrode loses control. The forward-blocking condition cannot be regained unless the gate signal is removed and the forward current falls below a certain low value called the holding, or sustaining, current. This is typically about 20 mA for a 50-A thyristor. In the forward conducting region, the thyristor has similar characteristics to those of a silicon diode of approximately the same current rating. Thus, the forward voltage drop is of the order of 1 to 1·5 V and increases slightly with current. Individual thyristor ratings have been increasing steadily over the years and units are now available with forward and reverse voltage ratings of 3000 V and current ratings of hundreds of amps.

For inverter applications, triggering by means of steep-fronted gate pulses is desirable in order to ensure that thyristors with different gate sensitivities turn on at precisely the same time. The gate pulse must not be removed until the anode current has increased to a value greater than the latching current which is somewhat greater than the holding current of the thyristor. With resistive loads, short gate pulses are satisfactory, but inductive loads delay the build-up of anode current and longer gate pulses are necessary. A medium-sized thyristor with an anode current rating of 50 A requires a gating signal of the order of 3 V and 100 mA. The turn-on time, which is the time taken from the instant the gate current is applied until the thyristor is fully conducting, is typically of the order of 0·5 to 4 microseconds.

2.2. THYRISTOR COMMUTATION

As already explained, the gate electrode loses control when the thyristor is switched into the conducting state. The process of interrupting current flow in a conducting thyristor is known as commutation, and the method used to achieve commutation is one of the main distinctions between inverter circuits.

When the thyristor is in the conducting state it contains a high concentration of positive and negative current carriers (holes and electrons). The simplest way to commutate the thyristor is to interrupt the current by a mechanical switch. The charges within the thyristor then recombine and it regains its forward-blocking ability. This is not feasible for high-frequency operation, and more effective static commutation methods have been developed. In these circuits, the thyristor current is interrupted by applying a reverse anode-to-cathode voltage across the thyristor. This reduces the forward current to zero and then produces a large pulse of reverse current for a very short interval of the order of a few microseconds. The reverse current sweeps the current carriers at the end-junctions, J_1 and J_3, out of the device. The anode current then falls to zero and the reverse-biased end-junctions are able to block the inverse voltage. However, the thyristor is not yet able to block forward voltage since current carriers are still present at junction J_2. A further time interval must be allowed for these carriers to recombine naturally. If forward voltage is re-applied before recombination is complete, the thyristor will immediately revert to the conducting state.

The turn-off time is defined as the time interval between the reduction of the anode current to zero and the regaining of the forward-blocking capability. The amount of forward current before commutation, influences the turn-off time, as the trapped charge at junction J_2 is directly proportional to this current. For present-day thyristors the turn-off time is typically between 3 and 100 microseconds.

Thyristor circuits may be classified in terms of the method used to achieve commutation.[58] Natural or phase commutation occurs in a.c.-operated circuits and will be examined in Section 2.4. In circuits which operate from a d.c. supply, the thyristor current must be forced to zero by the application of a reverse anode-to-cathode voltage which is produced by auxiliary components in the circuit. This process is called forced commutation and the extra equipment necessary is called the commutating circuit.

2.2.1. *Forced Commutation*

One method of applying the reverse voltage necessary for forced commutation is to switch a previously charged capacitor across the conducting thyristor (Fig. 2.3). The switch S represents a transistor or auxiliary thyristor which is triggered at the desired commutation instant. The current and voltage waveforms for the main thyristor are shown in Fig. 2.3(b). Before the commutation is initiated, the anode-to-cathode voltage has a value of approximately 1 V corresponding to the forward voltage drop in the conducting state. Following the switching of the charged capacitor across the thyristor there is a period during which the forward current drops to zero and a brief pulse of reverse current flows. This reverse current, which is limited by the stray inductance of the circuit, removes the free charge from the end-junctions with a corresponding reduction in the charge on the capacitor. This is known as the carrier storage effect. The drop in capacitor voltage is small, however, and its voltage is then blocked by the two end-junctions and appears as a reverse bias on the thyristor.

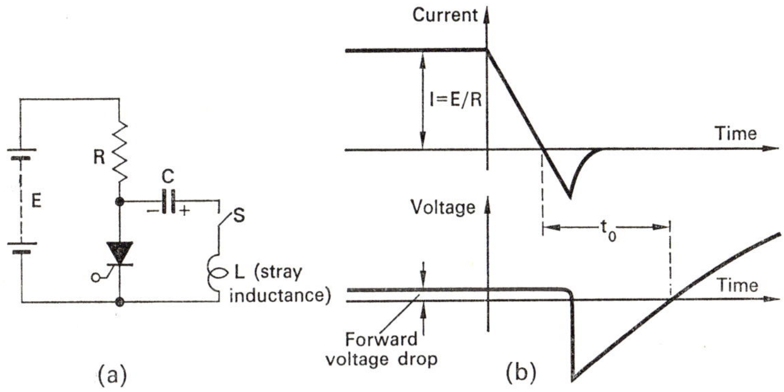

FIG. 2.3. Forced commutation by means of a charged capacitor: (a) basic circuit; (b) current and voltage waveforms.

The d.c. supply now charges up the capacitor with the opposite polarity. When its voltage reverses, the thyristor is again subjected to a forward voltage, but it will not conduct, provided recombination is complete at junction J_2 and it has, therefore, regained its forward-blocking capability. The time between the reduction of the anode current to zero and the re-application of forward voltage is called the circuit turn-off time or commutating time, t_o. This interval must be greater than the turn-off time required by the thyristor or a commutation failure will result. In practice, the thyristor turn-off time is measured by reducing the capacitance C, thereby reducing the commutating time provided by the circuit, until a commutation failure occurs.

The thyristor current before commutation determines the amount of free charge at junction J_2 and hence determines the recovery time required by the thyristor. The commutating circuit must provide sufficient turn-off time for the largest current to be commutated, and this determines the capacitance required. In a forced-commutated inverter, the thyristor has a significant advantage over the thyratron with its long de-ionization time. The use of thyristors permits a considerable reduction in the size and weight of the commutating capacitors. Further improvement is obtained by using inverter-type thyristors with guaranteed maximum turn-off times of 10 microseconds, or less.

2.3. THE FORCED-COMMUTATED SINGLE-PHASE BRIDGE INVERTER

A single-phase bridge inverter of the type used in many d.c. link converter circuits is shown in Fig. 2.4. The gating and commutating circuits are omitted for clarity. Four thyristors are connected in a bridge circuit with four diodes across them in inverse parallel. These feedback diodes are necessary to provide a path for the return of reactive energy, from inductive or capacitive loads, to the d.c. supply.

The bridge circuit may be divided into two sections, the left-hand section connected to load terminal A and the right-hand section with load terminal B. It is clear from the circuit that the two series thyristors in the same section or leg of the inverter must never conduct simultaneously as this would short-circuit the d.c. supply. The commutating circuit must ensure that a thyristor is always turned off before its series partner is gated on. Practical circuits which perform this function are described in Chapter 3.

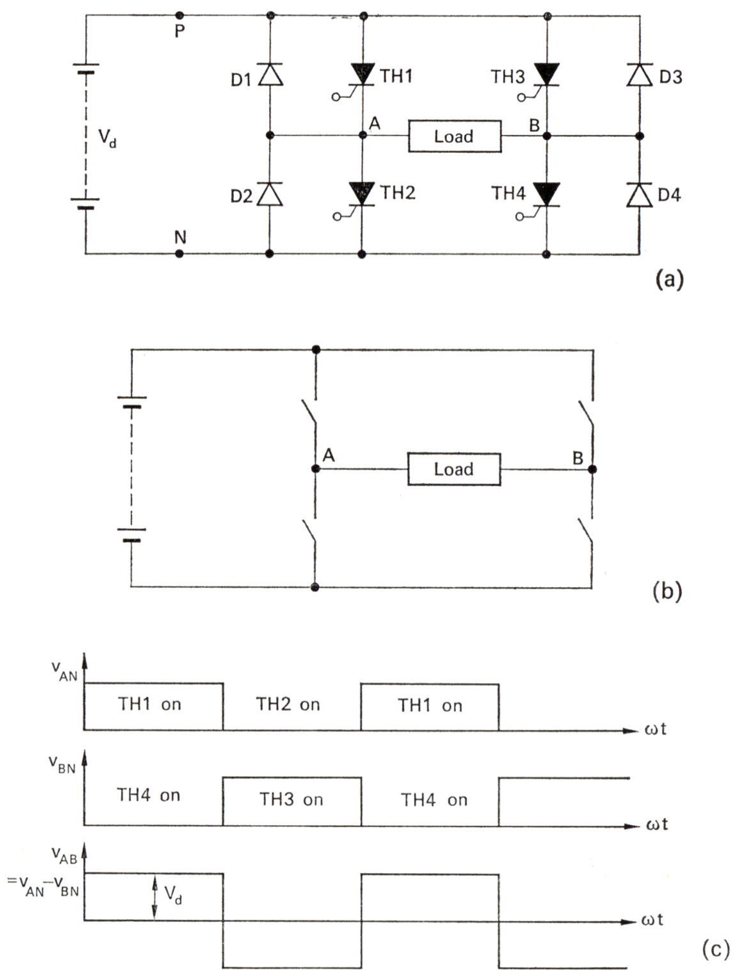

FIG. 2.4. Single-phase bridge inverter: (a) basic circuit; (b) equivalent circuit using mechanical switches; (c) voltage waveforms.

Thyristors TH1 and TH2 of the bridge circuit are gated and commutated so that they conduct for alternate 180° intervals. Neglecting the forward voltage drop of a conducting thyristor, terminal A is therefore connected alternately to the positive and negative sides of the d.c. supply. The thyristor circuit is equivalent to the mechanical switching system of

Fig. 2.4(b) where a conducting thyristor corresponds to a closed switch. If the commutation process occupies a negligible fraction of the half-cycle, the voltage of terminal A relative to the negative side of the d.c. supply has the square waveform, v_{AN}, of Fig. 2.4(c). Thyristors TH3 and TH4 also conduct alternately and terminal B is connected at any instant to either the positive or negative supply line. When TH1 and TH4 conduct simultaneously, the direct voltage input V_d is applied across the load. Similarly, when TH2 and TH3 conduct, the supply voltage is applied across the load in the reverse direction. If TH1 and TH4 are switched on and off simultaneously, a square waveform output voltage, v_{AB}, appears across the load as shown in Fig. 2.4(c). If the load is purely resistive, the current has an identical waveform. When supplying an inductive load such as an a.c. motor, the load current lags the applied voltage. Thus, when TH1 and TH4 are turned off, the load current continues to flow in the same direction for a time. During this period the current circulates through feedback diodes D2, D3 and the supply, and reactive energy is returned to the d.c. source. The feedback diodes also prevent the alternating load voltage from exceeding the direct input voltage, and a square voltage waveform of constant amplitude is generated at all frequencies and for all loading conditions. Feedback diodes should also be inserted in Fig. 2.4(b) to make the mechanical switching circuit equivalent to the thyristor circuit on inductive loads.

Three of the single-phase bridge circuits may be used to produce a three-phase output by maintaining the correct phase relationship between the gating signals to each circuit. The output frequency is altered by changing the frequency of the gating pulses.

2.3.1. *Simple Pulse-width Voltage Control*

In many applications it is necessary to vary the fundamental component of the alternating voltage output of the inverter. In the single-phase bridge circuit this may be achieved by varying the direct voltage input, V_d, and preserving the square waveform output voltage. An alternative method is to hold the supply voltage constant and vary the fundamental a.c. output voltage by using pulse-width modulation (pwm) techniques.

The simplest mode of pulse-width modulation is obtained by altering the phase of the conduction intervals of TH3 and TH4 with respect to TH1 and TH2. In Fig. 2.5 the gating of TH3 and TH4 is advanced by an angle γ equal to 120°. This means that at certain times

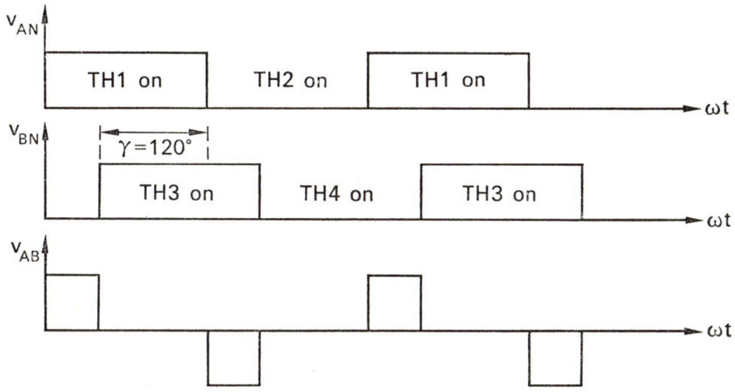

FIG. 2.5. Voltage control by simple pulse-width modulation.

in each cycle TH1 and TH3 conduct simultaneously, thereby connecting load terminals A and B to the positive supply line and producing an interval of zero output voltage. Similar considerations apply when TH2 and TH4 are conducting simultaneously. The resulting load voltage has the waveshape shown, with a pulse width of 60°. By advancing the gating of TH3 and TH4 from zero to 180°, the fundamental output voltage may be reduced from its maximum value to zero.

Certain values of γ also eliminate troublesome harmonics.[59] Thus, when γ equals 60°, a voltage waveform is obtained which has a zero voltage or "dwell" time of 60° and contains no third harmonic or multiples thereof. Similarly, when γ is 72°, the fifth harmonic is eliminated. However, for large values of γ, the output waveform consists of a series of very narrow pulses and the total harmonic content is excessive for many applications.

2.4. PHASE COMMUTATION IN SINGLE-PHASE A.C. CIRCUITS

In a thyristor circuit operated from an a.c. source, the natural reversal of supply voltage every half-cycle can be used to commutate a conducting thyristor, and no auxiliary apparatus is required. This process is called natural commutation or phase commutation.

2.4.1. *Single-phase Half-wave Rectification*

The simplest thyristor circuit using phase commutation is the single-phase half-wave rectifier. Figure 2.6 shows the circuit with a resistive load. The thyristor blocks current flow in both directions until it is gated. If the gating pulse is applied during the positive half-cycle of anode voltage, the thyristor immediately conducts and the supply voltage appears across

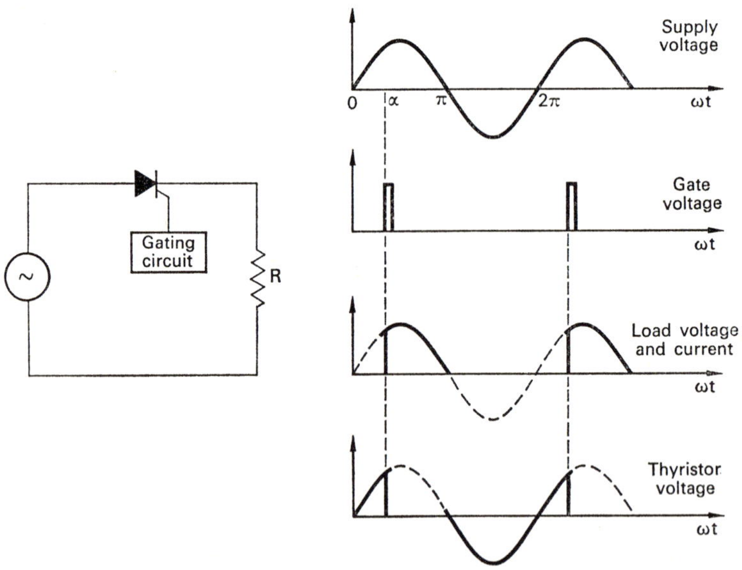

FIG. 2.6. Single-phase half-wave rectifier with resistive load.

the load for the remainder of the positive half-cycle. This assumes that the forward voltage drop of the thyristor is negligible compared with the supply voltage, as is usually the case. The load voltage and current waveforms are identical in the case of a purely resistive load, and at the end of the positive half-cycle the thyristor current falls to zero, and the supply voltage reverses. This applies a reverse bias across the thyristor and rapidly turns it off. The brief pulse of reverse current, which flows for a few microseconds, is ignored in Fig. 2.6. The thyristor is now reverse-biased for the complete negative half-cycle of the supply voltage and this provides more than adequate recovery time unless the supply frequency is increased to tens of kHz.

When a regular diode is used in this circuit the full positive half-cycle of supply voltage appears across the load. The angle by which the thyristor retards the start of conduction is called the firing angle or delay angle, α. Variation of α from 0° to 180° reduces the average d.c. output voltage from a maximum value to zero. This process is known as phase control.

When the load is inductive, the current builds up gradually at the start of conduction, and subsequently the decaying inductive current holds the thyristor in conduction after $\omega t = \pi$. The negative supply voltage then opposes the inductive current, reducing it to zero, and the thyristor turns off and is reverse-biased for the remainder of the negative half-cycle. The gating pulse has a rectangular waveform and lasts from $\omega t = \alpha$ to π. The sustained pulse allows the inductive load current to increase above the latching value of the thyristor.

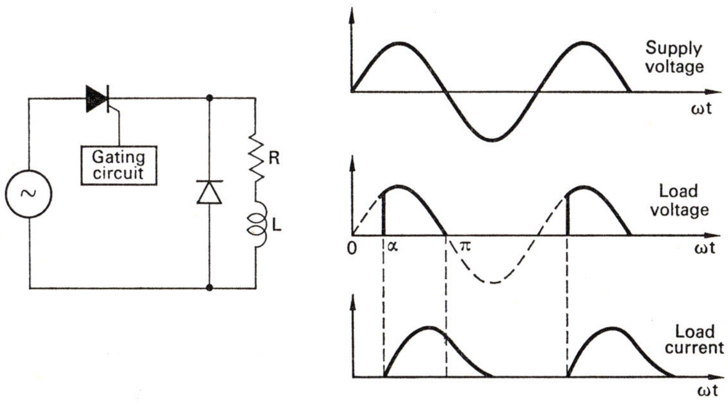

FIG. 2.7. Single-phase half-wave rectifier with inductive load and free-wheeling diode.

In practice, the half-wave circuit is usually modified for inductive loads by placing a "free-wheeling" diode across the load as shown in Fig. 2.7. When the thyristor is triggered, the current builds up as before but, at the end of the positive half-cycle, the negative supply voltage is applied directly across the thyristor, and the current in the inductive load decays naturally by circulating through the free-wheeling diode. In this circuit, the load voltage waveform for a given firing angle, α, is independent of the time constant of the load (Fig. 2.7).

A free-wheeling diode is often placed across inductive loads in thyristor and transistor switching circuits. By allowing the inductive current to decay at its natural time constant, it is possible to avoid the destructive overvoltages which occur when the current flow is suddenly interrupted.

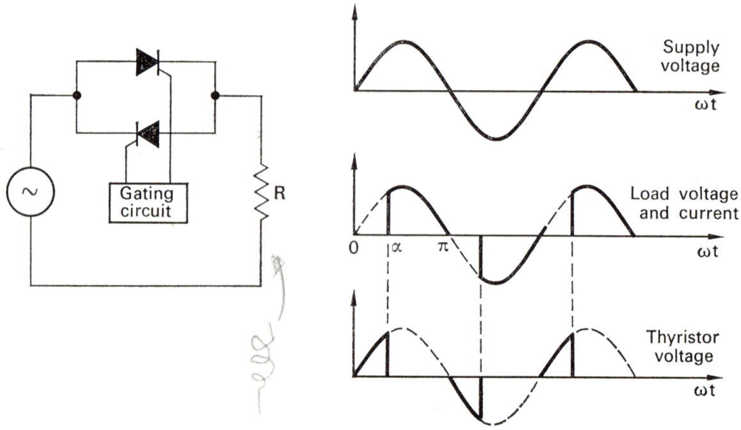

FIG. 2.8. Phase control of alternating voltage in a single-phase back-to-back thyristor circuit with a resistive load.

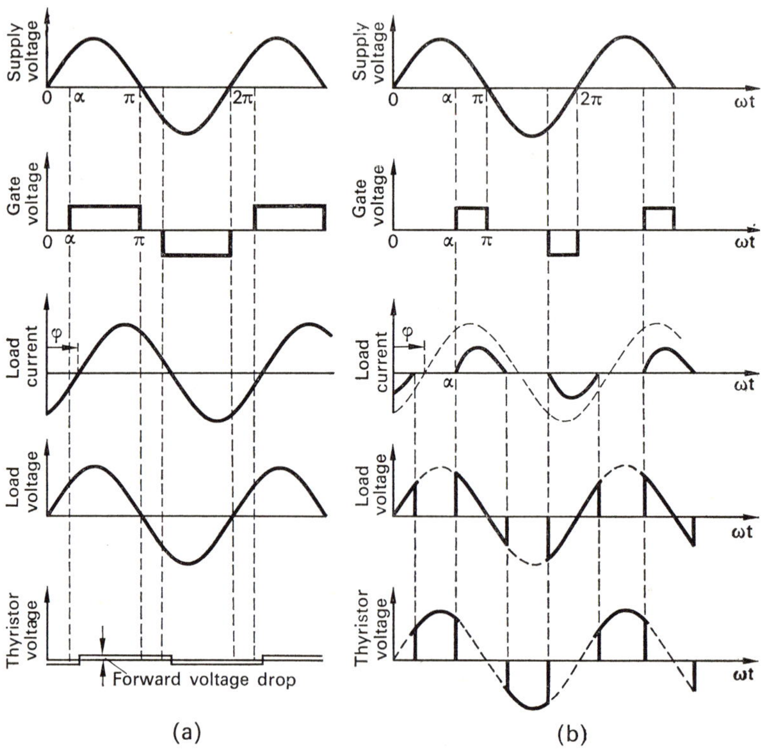

FIG. 2.9. Voltage and current waveforms for a single-phase back-to-back thyristor circuit with an inductive load: (a) for a firing angle α which is less than ϕ, the phase angle of the load; (b) for α greater than ϕ.

2.4.2. *A.C. Power Control*

Full-wave phase control can be obtained if a second thyristor is introduced into the half-wave circuit as shown in Fig. 2.8. The two thyristors are connected in inverse parallel or "back-to-back" and are triggered alternately at identical points in their anode-to-cathode voltage cycles. A symmetrical alternating voltage is, therefore, applied across the load. This voltage waveform has no d.c. component but odd harmonics are present, principally the third. For a purely resistive load, the voltage and current waveforms are as shown in Fig. 2.8. With an inductive load, current flow is again maintained by the induced e.m.f. for an interval after the supply voltage has reversed its polarity. The duration of this interval is determined by the load power factor and typical waveforms are shown in Fig. 2.9. If the phase angle of the inductive load, for normal sinusoidal operation, is denoted by ϕ, then the load current and voltage are sinusoidal for any firing delay α which is less than ϕ, because the incoming thyristor is already triggered before the natural reversal of current direction takes place. However, if α exceeds ϕ, the current flow becomes discontinuous (Fig. 2.9) and the pulses of load current are non-sinusoidal. The current pulses decrease in amplitude as α is increased, and zero load current is obtained when $\alpha = \pi$. The precise mathematical expression for the current waveform is obtained in the usual manner by solving the differential equation for the inductive circuit and substituting the appropriate boundary conditions.[60]

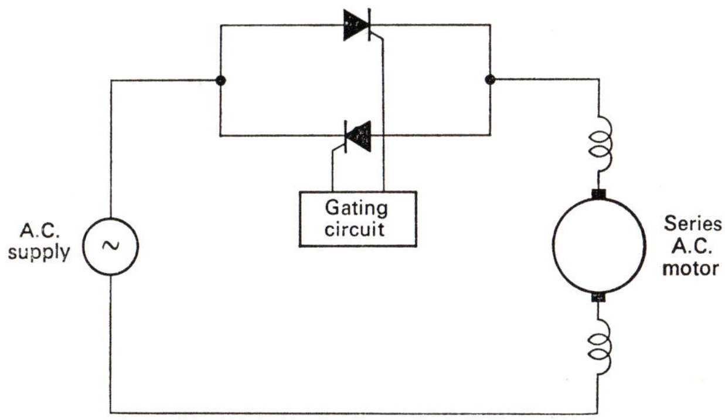

FIG. 2.10. Basic circuit for speed control of a single-phase a.c. series motor.

Variation of the thyristor delay angle α controls the average power delivered to the load by the a.c. supply. A bi-directional thyristor or Triac may be used in place of the two back-to-back thyristors. These circuits can be used as power regulators for heating or lighting loads. They are also widely used for speed control of single-phase a.c. series or universal motors.[61] A typical motor control scheme is shown in Fig. 2.10. Three-phase versions of the circuit are studied in Chapter 9.

2.5. PHASE-CONTROLLED RECTIFICATION AND INVERSION IN POLYPHASE A.C. CIRCUITS[62,63]

The single-phase rectifier has a pulsating output voltage and poor transformer utilization. Polyphase circuits are therefore used in high-power industrial applications. For simplicity, the operation of a three-phase half-wave circuit is analysed. The three thyristors have a

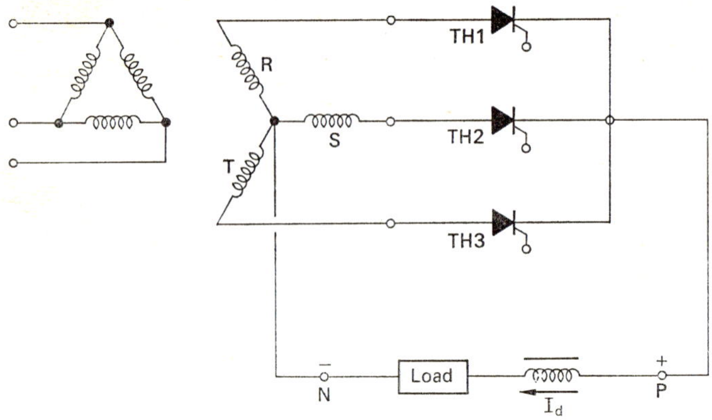

FIG. 2.11. The three-phase half-wave rectifier circuit.

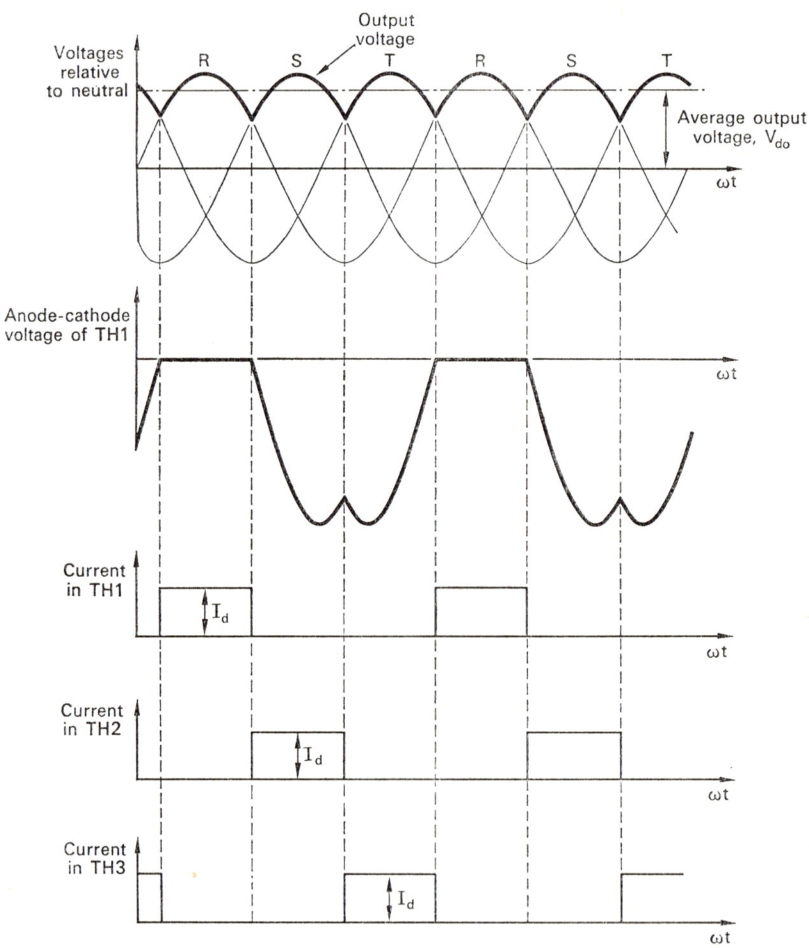

FIG. 2.12. Voltage and current waveforms for a three-phase half-wave rectifier with a highly inductive load and zero firing angle.

common cathode connection, as in Fig. 2.11, and this forms the positive terminal of the rectified d.c. output. The transformer neutral constitutes the negative output terminal.

In these multiphase a.c. circuits, the natural commutation process consists of a cyclic transfer of current from one thyristor to the next. Assume, for the moment, that the thyristors are replaced by regular diodes and hence no firing delay is possible. Each diode then conducts the full output current for one-third of a cycle while its anode voltage is the most positive of the three. Figure 2.12 shows the transformer (or anode) voltages relative to neutral and the rectified output voltage assuming zero forward voltage drop. In effect, the positive load terminal P is automatically connected by a conducting diode to whichever phase of the secondary is most positive, and the non-conducting diodes are reverse-biased relative to the common cathode terminal. These waveforms are also obtained using thyristors, provided each one is triggered as soon as its anode-to-cathode voltage becomes positive.

When feeding a resistive load, the output current waveform is a replica of the output voltage. If a large inductor is connected in the d.c. circuit, it will suppress harmonic currents and maintain a steady output current, I_d. Each thyristor carries this current for one-third of the cycle, as shown in Fig. 2.12.

2.5.1. *Delayed Commutation*

The average output voltage is controlled by using a thyristor circuit with delayed commutation. The firing angle or delay angle, α, is measured from the point where commutation occurs when regular diodes are used. Figure 2.13 shows the voltage and current waveforms for an inductive load which maintains constant current. The gating pulses are now delayed and each thyristor blocks the positive anode voltage for an angle α. This forces the preceding phase to continue supplying the load current although it is not the most positive. When the incoming thyristor is triggered, the load current immediately transfers to it, and a reverse bias is applied across the outgoing thyristor, causing it to turn off. The output voltage now has the waveform shown in Fig. 2.13 and the average value has obviously been reduced.

A general expression for the average output voltage is obtained as follows: assume an m-phase half-wave circuit in which each phase conducts for $2\pi/m$ radians in one cycle of the supply voltage. The instantaneous phase voltage is given by $v = \sqrt{2}\, V_{ph} \cos \omega t$ where V_{ph} is the r.m.s. value of the transformer secondary voltage and the time origin is taken at the instant of peak voltage (Fig. 2.14). If the firing angle is zero, conduction takes place from $(-\pi/m)$ to $(+\pi/m)$. If the firing angle is α, the phase conducts from $[(-\pi/m) + \alpha]$ to $[(+\pi/m) + \alpha]$, and the average output voltage, V_d, equals the average height of the shaded area in Fig. 2.14.

$$\therefore\ V_d = \frac{m}{2\pi} \int_{[-(\pi/m)+\alpha]}^{[(\pi/m)+\alpha]} \sqrt{2}\, V_{ph} \cos \omega t \, d(\omega t) = \sqrt{2}\, V_{ph} \left(\frac{m}{\pi}\right) \sin\left(\frac{\pi}{m}\right) \cos \alpha .$$

When the delay angle is zero, V_d has a maximum value

$$V_{do} = \sqrt{2}\, V_{ph} \left(\frac{m}{\pi}\right) \sin\left(\frac{\pi}{m}\right). \tag{2.1}$$

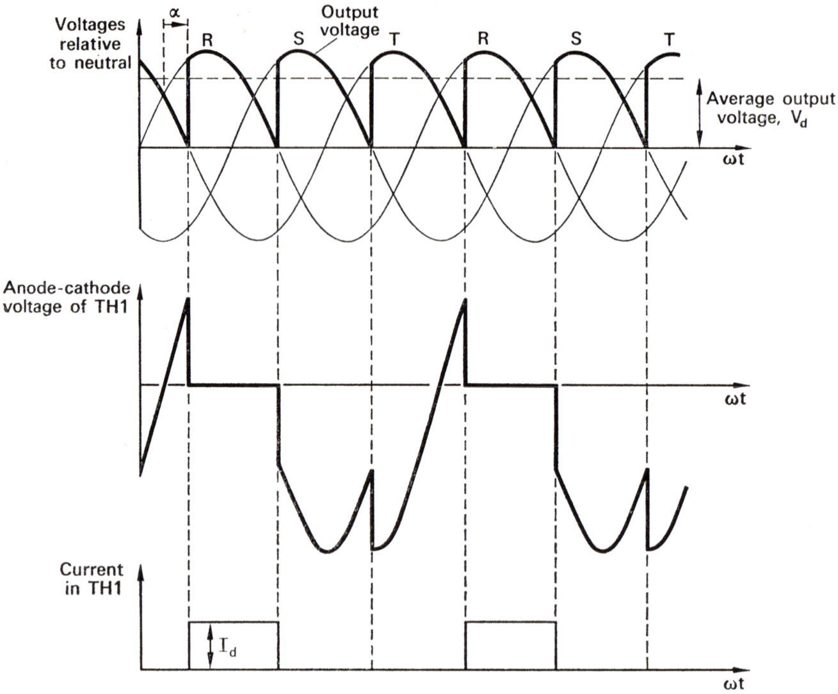

FIG. 2.13. Voltage and current waveforms for a three-phase half-wave rectifier with a highly inductive load and delayed commutation.

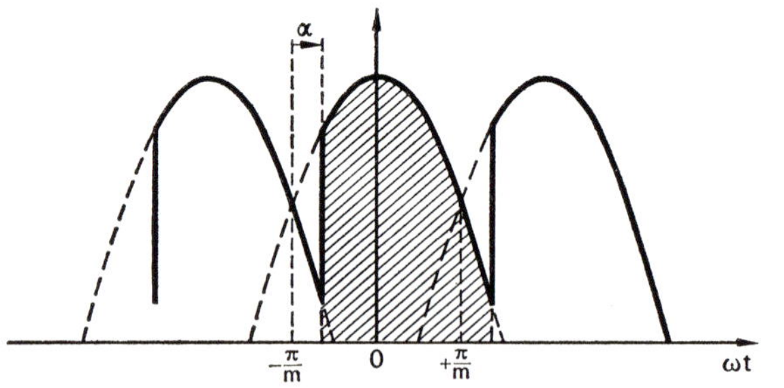

FIG. 2.14. Output voltage waveform for an m-phase half-wave rectifier with a firing angle α.

The mean output voltage, with phase control, is therefore given by

$$V_d = V_{do} \cos \alpha . \tag{2.2}$$

This formula assumes continuous current conduction and is not valid if current flow becomes discontinuous as a result of delayed commutation. The output voltage and current waveforms for a three-phase half-wave rectifier with a firing delay of 60° are shown in Fig. 2.15(a) for a highly inductive load and in Fig. 2.15(b) for a resistive load. When the d.c. load is highly inductive it tends to maintain constant output current even with large delay angles, and the thyristor current retains its rectangular waveform. Consequently, the induced e.m.f. in the d.c. inductor maintains current flow even when the anode polarity has reversed. This means that energy is being returned from the magnetic field of the inductor through the transformer to the supply, and the circuit is temporarily acting as an inverter. However, with a firing delay of 60° the mean output voltage, V_d, is positive, and there is a net power flow from the a.c. supply to the load. With a delay of 90°, V_d is zero and energy oscillates between the inductor and the supply, with no net power flow (assuming zero circuit losses). If the firing is delayed beyond 90°, V_d is negative and inverter operation may be obtained by introducing a d.c. source in series with the inductor.

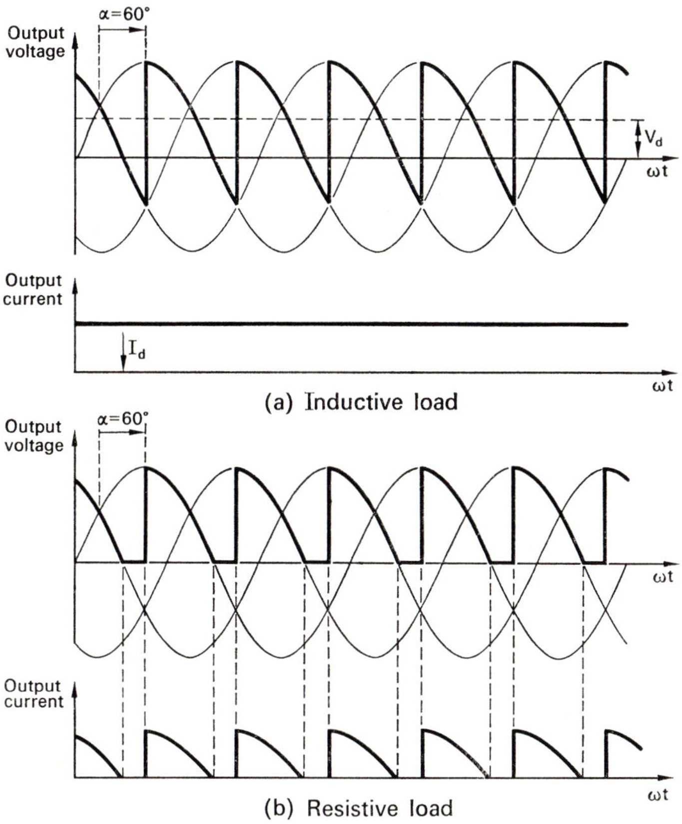

FIG. 2.15. Output voltage and current waveforms when commutation is delayed by 60°: (a) with a highly inductive load; (b) with a resistive load.

In the case of a resistive load, each thyristor ceases to conduct as soon as its anode voltage goes negative. In general, this causes discontinuous current flow in half-wave circuits when the delay angle exceeds $(\pi/2 - \pi/m)$, i.e. 30° in the case of a three-phase circuit. For delay angles less than this, the formula $V_d = V_{do} \cos \alpha$ is valid, but for greater delays the formula is

$$V_d = V_{do} \left\{ \frac{1 - \sin(\alpha - \pi/m)}{2 \sin(\pi/m)} \right\}. \tag{2.3}$$

2.5.2. *Overlap*

In the analysis to date, it has been assumed that the load current I_d commutates instantaneously from one thyristor to the next. In practice, the supply transformer has leakage inductance which prevents the instantaneous transfer of current. As a result, there is a commutation period or angle of overlap, u, during which two thyristors conduct simultaneously and two transformer phases are short-circuited. The voltage difference between the phases circulates a commutating current which reduces the current in the outgoing thyristor to zero and increases the current in the incoming thyristor to I_d (Fig. 2.16). In a phase-controlled rectifier, the overlap angle u decreases as the delay angle α is increased. This is caused by the increased commutating voltage available at large delay angles.

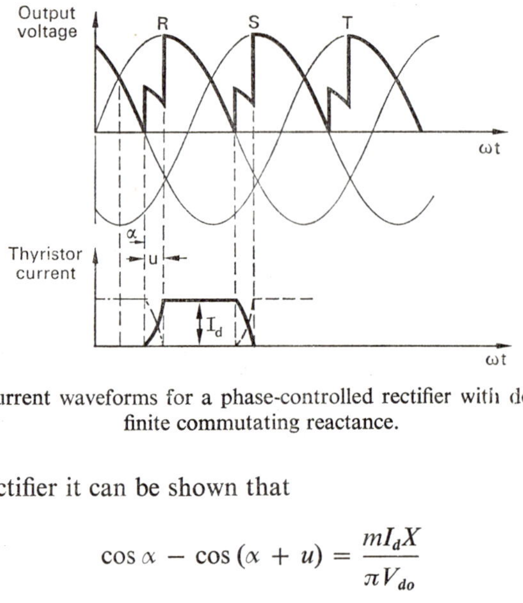

FIG. 2.16. Voltage and current waveforms for a phase-controlled rectifier with delayed commutation and finite commutating reactance.

For a half-wave rectifier it can be shown that

$$\cos \alpha - \cos (\alpha + u) = \frac{mI_dX}{\pi V_{do}} \tag{2.4}$$

where X is the equivalent leakage reactance or commutating reactance per phase.

During the overlap period, the positive load terminal is connected simultaneously to two phases and it assumes a potential equal to the average value of the two conducting phases. When the incoming thyristor is triggered the output voltage jumps to this average value and remains there until commutation is completed, when it jumps to the incoming phase voltage (Fig. 2.16). The average output voltage, V_d, is therefore below the theoretical value obtained with instantaneous commutation. It is readily shown that the voltage drop due to overlap is independent of the delay angle α and is given by $(mI_dX)/2\pi$. In general, therefore, the mean output voltage of a half-wave rectifier with firing delay and overlap is

$$V_d = V_{do} \cos \alpha - \frac{mI_dX}{2\pi} \tag{2.5}$$

where V_{do} is the output voltage with zero delay and no overlap. Thus, for a fixed delay angle, α, the output voltage decreases linearly with load current.

2.5.3. *The Phase-controlled Inverter*

In a d.c. circuit, a reversal in the direction of power flow is normally associated with a reversal of the current direction, but the same effect is obtained if the voltage polarity is reversed and the current direction is maintained. The latter technique must be used in the phase-controlled inverter as it is a unidirectional current device.

In a normal rectifier, each phase of the a.c. supply delivers current through a thyristor while the phase or anode voltage is positive. Thyristor current must flow from anode to cathode and the direction of energy flow can only be reversed if the a.c. supply delivers current during periods of negative anode voltage. This is achieved by delaying commutation until the anode voltage goes negative and then forcing current flow in opposition to the negative phase voltage. The d.c. source which produces this current delivers energy to the a.c. supply and inverter operation is achieved.

For inversion, the commutation instant must be delayed beyond 90° when the mean output voltage $V_d = V_{do} \cos \alpha$ becomes negative. Figure 2.17 shows the voltage and current waveforms, assuming a constant current I_d is maintained by means of the d.c. inductor. The transformer neutral is now the positive d.c. terminal and the d.c. supply voltage V exceeds

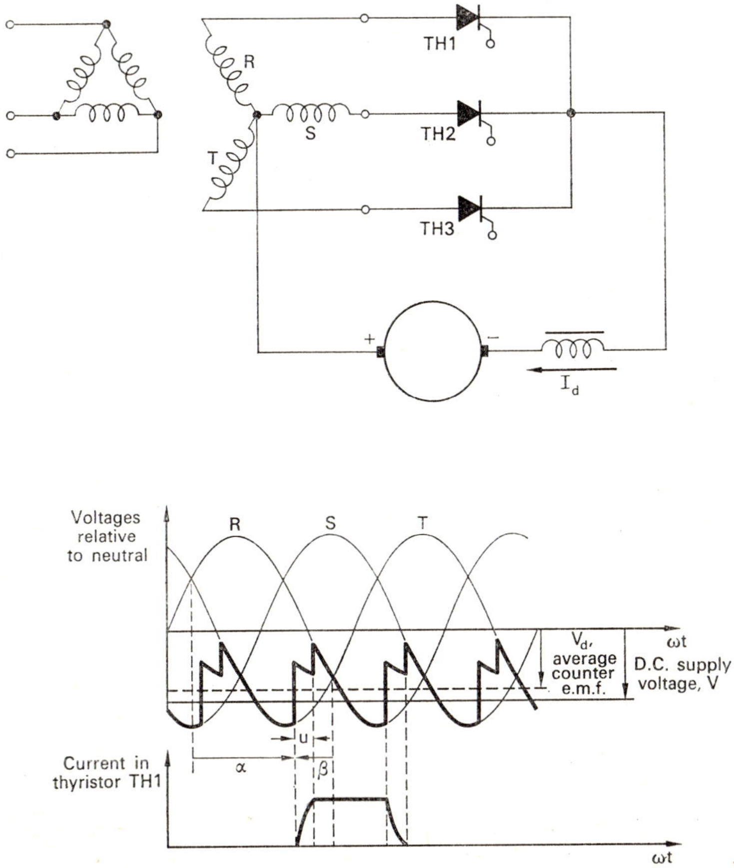

FIG. 2.17. Basic circuit diagram and waveforms for the phase-controlled inverter.

the average counter e.m.f. V_d by the amount necessary to circulate the current I_d through the circuit resistance. Phase commutation is obtained because the incoming thyristor is triggered while its anode is more positive than the anode of the conducting thyristor. The voltage difference between phases circulates a commutating current in the usual manner. If the delay angle α is greater than π, this commutating voltage is no longer available and conduction continues into the positive half-cycle. The rectified voltage then aids the d.c. supply voltage and a short-circuit condition is produced. Commutation must, therefore, be completed before the instant at which the phase voltages become equal. The angle by which firing is advanced ahead of this point is denoted by β, where $\beta = \pi - \alpha$. The angle β must always be large enough to allow sufficient time for overlap and turn-off of the thyristor. For inverter operation, the voltage drop due to overlap has the effect of increasing the negative back e.m.f. of the inverter, which therefore increases with load current.

Figure 2.18 shows the anode–cathode voltage of a thyristor in the inverter circuit, assuming negligible overlap. After the transfer of current, a reverse bias is applied to the outgoing thyristor for a short interval. This interval is called the commutation margin and must be greater than the thyristor recovery time or a commutation failure will occur. Note the

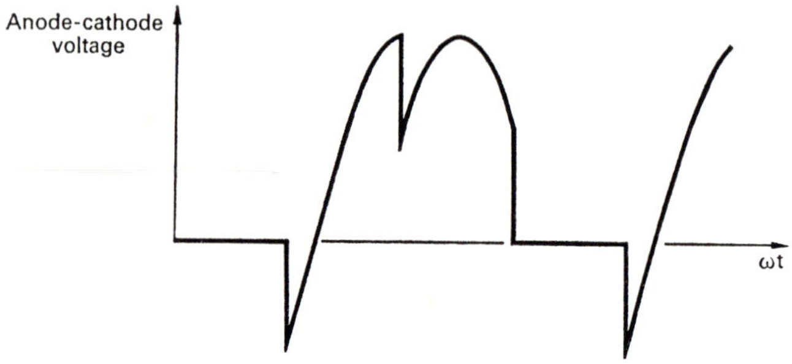

FIG. 2.18. Thyristor anode-cathode voltage in a phase-controlled inverter with zero commutating reactance.

similarity between the thyristor voltage waveforms of Fig. 2.18 for phase commutation and Fig. 2.3(b) for forced commutation. However, the phase-controlled inverter, unlike the forced-commutated inverter, needs no auxiliary commutating apparatus since the commutating voltages are provided by the a.c. supply into which the inverter is feeding. This principle is used in a.c./d.c./a.c. power transmission systems and is also the basic commutation process in the cycloconverter (Chapter 4).

2.5.4. *Reactive Power and Power Factor*

In a phase-controlled rectifier, the introduction of a firing delay causes the thyristor current to lag the anode or phase voltage (Fig. 2.16). If the load is highly inductive and overlap is negligible, the thyristor current retains its rectangular waveform but is delayed by an angle α. The displacement ϕ between the phase voltage and the fundamental of the phase current is equal to the delay angle, α, and hence the fundamental power factor, $\cos\phi$, decreases in proportion to the per-unit output voltage V_d/V_{do}. This phase lag in transformer secondary current is reflected on the primary side, resulting in a low input power factor. The presence of overlap tends to increase the phase lag of the fundamental current and even with zero firing delay the input power factor is somewhat less than unity.

In the case of an inverter, the fundamental component of output current would have unity power factor if commutation was instantaneous and could be delayed until $\alpha = \pi$. In practice, the firing must be advanced in order to allow sufficient time for overlap and turn-off, and hence the output current always leads the phase voltage. A phase-controlled inverter, therefore, delivers current to the supply at a leading power factor and the inverter may be regarded as a generator of leading reactive power or a consumer of lagging reactive power. The inverter delivers real power to the a.c. supply but its lagging reactive power requirements must be supplied by the synchronous alternators of the a.c. system into which the inverter is feeding. Alternatively, compensating capacitors may be connected on the a.c. side of the inverter in order to reduce the reactive power consumption.

In general, when a thyristor converter is connected to the a.c. supply, a discontinuous non-sinusoidal current flows in the a.c. system and one must distinguish between the real power, the reactive power and the harmonic power.

The real power per phase, P, is equal to the product of the r.m.s. phase voltage V_{ph} and the in-phase component of the fundamental current. If I_1 is the fundamental r.m.s. current and $\cos\phi$ is the fundamental power factor,

$$P = V_{ph} I_1 \cos\phi.$$

The reactive power, Q, as usual, is the product of the r.m.s. phase voltage and the quadrature component of fundamental current.

$$\therefore Q = V_{ph} I_1 \sin\phi.$$

The harmonic power, Q_{har}, is given by the product of the phase voltage V_{ph} and the total r.m.s. harmonic current I_{har}.

$$\therefore Q_{har} = V_{ph} I_{har}.$$

The harmonic power, like the reactive power, represents real power oscillating between the load and the supply.

If I_{ph} is the total r.m.s. current, including harmonics, then

$$I_{ph} = \sqrt{(I_1^2 + I_{har}^2)}$$

and the total apparent power, S, is given by

$$S = V_{ph} I_{ph} = \sqrt{(P^2 + Q^2 + Q_{har}^2)}.$$

Strictly speaking, when harmonics are present, the fundamental power factor, $\cos\phi$, is called the displacement factor and the ratio of real to apparent power is called the total power factor, λ.

Thus

$$\lambda = \frac{P}{S} = \frac{V_{ph} I_1 \cos\phi}{V_{ph} I_{ph}} = \frac{I_1}{I_{ph}} \cos\phi$$

or

$$\lambda = \mu \cos\phi \qquad (2.6)$$

where μ is the ratio of the fundamental r.m.s. current to the total r.m.s. current and is called the distortion factor.

2.5.5. *Free-wheeling Operation*

The operation of a polyphase a.c. rectifier with an inductive load is modified when a free-wheeling diode is connected across the load (Fig. 2.19). As in the single-phase circuit of Fig. 2.7, this diode blocks the flow of rectifier current, but allows the load current to circulate during periods of negative voltage. A continuous load current is therefore obtained at

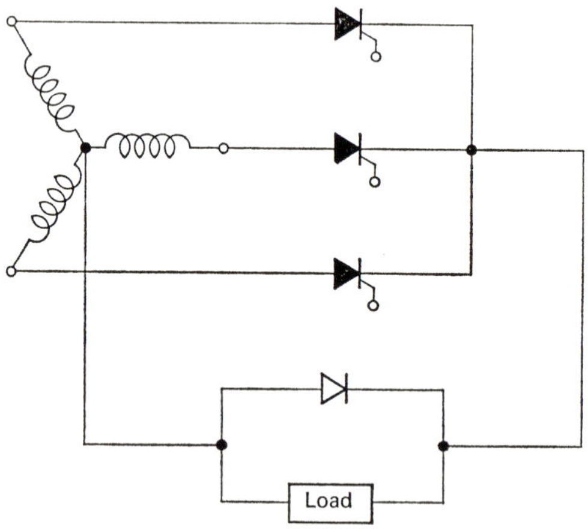

FIG. 2.19. The three-phase half-wave rectifier with a free-wheeling diode.

large delay angles, even though the rectifier current is discontinuous. When the free-wheeling diode is removed from the polyphase circuit, the inductive load maintains a continuous current and forces rectifier current flow during periods of negative output voltage. In the modified circuit, the free-wheeling diode immediately comes into conduction when the rectified voltage reverses its polarity. The output current is then maintained by the load e.m.f., and the rectifier current is zero. The inductive load is effectively short-circuited by the conducting diode until the rectified voltage again becomes positive. The load voltage therefore has the same discontinuous waveform as in Fig. 2.15(b) for a normal three-phase rectifier supplying a resistive load. The presence of the free-wheeling diode prevents a reversal in the polarity of the output voltage and so reduces the ripple content of the output voltage and current waveforms. It also modifies the alternating current waveform of the phase-controlled rectifier so that the reactive power consumption is reduced and the fundamental power factor is improved.(63) The free-wheeling diode must be removed if inverter operation is required.

2.6. THE PHASE-CONTROLLED THREE-PHASE BRIDGE RECTIFIER

The basic principles of phase-controlled rectification and inversion have now been introduced. In practice, a three-phase half-wave circuit would use a rectifier transformer with a zig-zag secondary connection.(62) This prevents magnetic saturation of the transformer core by the d.c. component of the secondary current. Other circuits, using interphase transformers and fork connections, have been developed in order to improve transformer utilization. However, the basic theory presented above is readily extended to these more complex configurations.

A circuit arrangement which is commonly used with thyristors is the three-phase bridge circuit of Fig. 2.20(a). This is a relatively simple circuit using six thyristors and the circuit is redrawn in Fig. 2.20(b) to emphasize that it may be regarded as two three-phase half-wave circuits connected in series. The upper group has a common cathode connection and

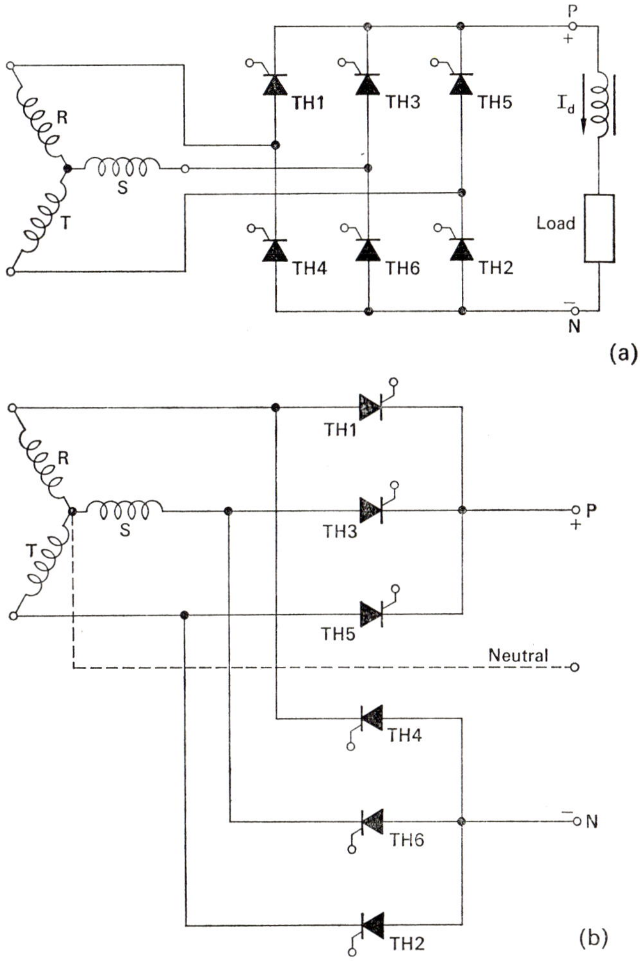

FIG. 2.20. The basic three-phase bridge rectifier: (a) the usual circuit diagram; (b) the bridge circuit redrawn as two half-wave circuits.

operates in the usual manner. If there is no firing delay, the thyristor with the most positive anode conducts and produces a positive output voltage at P relative to the neutral. The lower half-wave group has a common anode connection and the thyristor with the most negative cathode conducts and makes terminal N negative with respect to the neutral. The potential of P therefore follows the upper envelope of the alternating input voltage while N follows the lower envelope (Fig. 2.21). The currents in the neutral connection cancel one another, and the neutral terminal is not used. The output voltage between the P and N terminals is the vertical distance between the upper and lower envelopes in Fig. 2.21, and hence it is formed by segments of the three-phase line-to-line voltages and has a six-phase ripple as shown. The average output voltage is twice that of a three-phase half-wave circuit. Thus, the mean output voltage with negligible overlap and zero firing delay is

$$V_{do} = \frac{3\sqrt{6}}{\pi} V_{ph} = 2.34 V_{ph} = 1.35 V_L \tag{2.7}$$

where V_{ph} is the r.m.s. phase voltage, and V_L is the r.m.s. line voltage.

Each thyristor carries the full direct current I_d for 120° in each cycle. Two thyristors always conduct simultaneously in order to complete the circuit, but the conduction periods

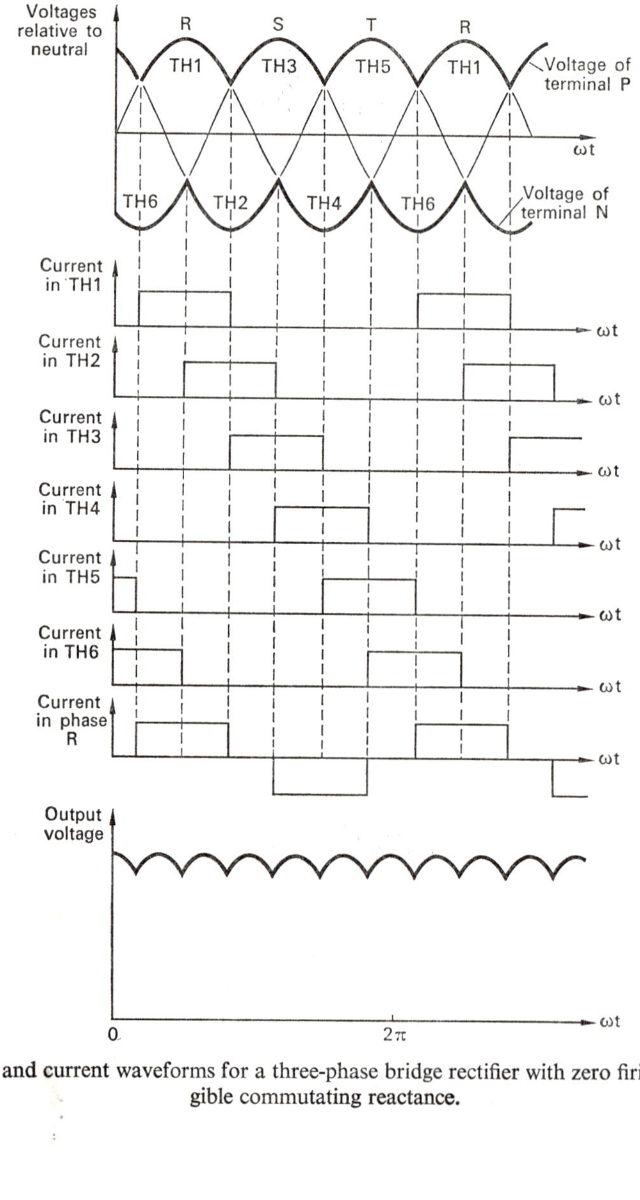

FIG. 2.21. Voltage and current waveforms for a three-phase bridge rectifier with zero firing angle and negligible commutating reactance.

of elements in the upper and lower groups do not coincide in time. For a phase sequence R, S, T, the thyristors conduct in pairs: TH1 and TH2; TH2 and TH3; TH3 and TH4; TH4 and TH5; TH5 and TH6, and they are consequently numbered in their correct firing sequence in Fig. 2.20. A commutation occurs every 60° and alternate commutations take place in the upper and lower groups. If the d.c. load has sufficient inductance to maintain a constant output current I_d, the thyristor and a.c. line currents are as shown in Fig. 2.21, and the line current has no d.c. component.

When a firing delay is introduced, the voltage waveform is modified and the mean output voltage is given by $V_d = V_{do} \cos \alpha$, as before. The presence of overlap causes further modification of the voltage waveform and introduces a voltage drop of $3I_dX/\pi$, which is twice that for a three-phase half-wave circuit. Inverter operation is again obtained when the firing angle is greater than 90° and all the previous inverter theory is directly applicable.

The three-phase bridge circuit is popular for thyristor applications because the thyristors are subjected to only half the peak reverse voltage ($1{\cdot}045V_d$) as compared with a six-phase half-wave circuit or double three-phase circuit with interphase transformer ($2{\cdot}09V_d$). For a certain direct current, I_d, the thyristors in the bridge circuit must carry twice as much current, but thyristors are more critical of voltage requirements and their cost increases sharply with voltage rating. In a bridge rectifier, the output voltage is reduced by the voltage drop of two conducting thyristors in series, but this is not normally excessive due to the small forward voltage drop of a thyristor. The bridge circuit has the advantage that there is no d.c. component in the a.c. current and a transformer supply is unnecessary. A neutral connection is not required and this is often an advantage in a.c. motor applications, since the star point of the machine winding may not be available.

2.6.1. *The Half-controlled Bridge Rectifier*

The fully controlled three-phase bridge rectifier has six thyristors connected as in Fig. 2.20(a). As explained above, this circuit operates as a rectifier when each thyristor has a firing angle α which is less than 90°, and functions as an inverter for α greater than 90°. If inverter operation is not required, the circuit may be simplified by replacing three controlled rectifiers by regular diodes as in Fig. 2.22. This is economically attractive since diodes are considerably less expensive than thyristors.

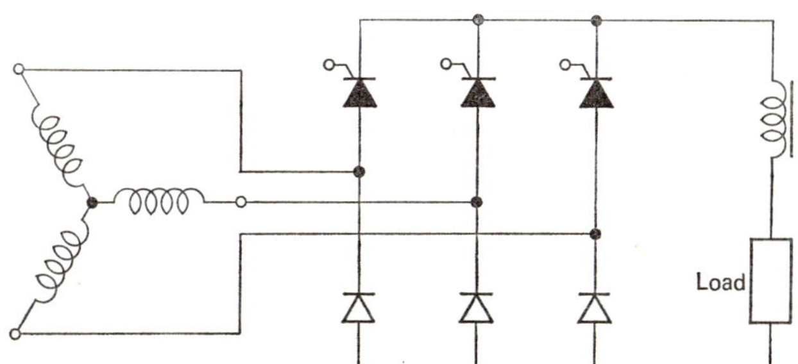

FIG. 2.22. The half-controlled three-phase bridge rectifier circuit.

Adopting the technique of Section 2.6, the half-controlled bridge is analysed by considering it as a phase-controlled half-wave circuit in series with an uncontrolled half-wave rectifier. The output voltage of the uncontrolled rectifier is constant, but the controlled rectifier delivers a variable output voltage which is added to that of the uncontrolled section. By delaying the thyristor firing beyond 90°, the controlled section operates as an inverter, and power delivered by the diodes is returned to the a.c. supply by the thyristors. The inverter back e.m.f. opposes the rectifier voltage and reduces the resultant output voltage of the bridge. Theoretically, zero output voltage is obtained when the firing angle is 180° but, in practice, the circuit may malfunction at large delay angles due to the failure of a thyristor to commutate in the limited time available. This is known as the "half-waving" effect.

The output voltage waveforms of a half-controlled bridge are similar to those of a fully controlled bridge with a free-wheeling diode. The output voltage ripple gradually changes its character from a six-phase ripple when α is zero, to a three-phase ripple when α is large. The input power factor is also higher than that of a fully controlled bridge rectifier without a free-wheeling diode.

CHAPTER 3

The D.C. Link Converter

3.1. INTRODUCTION

The d.c. link converter is a two-stage conversion device in which power from the a.c. network is first rectified to d.c. and then inverted to a.c. at variable frequency. Rectification is performed by conventional three-phase bridge circuits, and consequently, in this chapter, attention is focused on the d.c.–a.c. inverter which delivers variable-frequency power to the three-phase a.c. motor. Voltage control is incorporated in the converter in order to vary the alternating voltage output and so maintain the correct magnetic conditions in the motor. The complexity of the converter circuitry is determined by the quality of the output voltage waveform. If the motor speed must be reduced rapidly, electrical braking is also incorporated.

3.2. THE BASIC THREE-PHASE, SIX-STEP INVERTER CIRCUIT

The static inverter must generate a three-phase output of variable frequency and variable voltage. The simplest three-phase inverter is the three-phase bridge inverter circuit of Fig. 3.1, in which the gating and commutating circuits have been omitted for clarity.

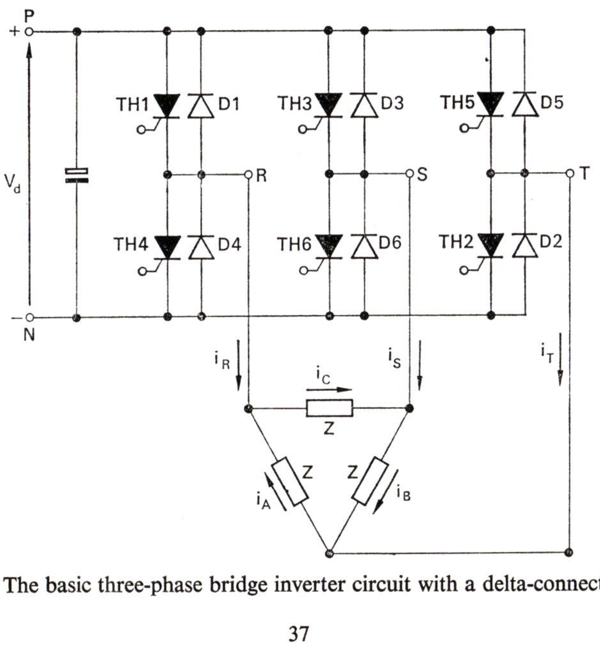

FIG. 3.1. The basic three-phase bridge inverter circuit with a delta-connected load.

Suitable commutating circuits will be described in Section 3.4, but the basic inverter operation is independent of the commutation method used. The inverter circuit shown is similar to the phase-controlled bridge circuit of Fig. 2.20(a), but forced commutation is now employed. Feedback diodes are included in the circuit and, when operating as a six-step inverter, each thyristor has a conduction period of 180° instead of the 120° conduction period of the phase-controlled circuit.

The operation of the six-step inverter is essentially the same as that of the single-phase bridge inverter of Fig. 2.4. A third leg is added to the single-phase circuit so that each output terminal R, S and T is connected alternately by a conducting thyristor to the positive and negative sides of the d.c. supply. A three-phase output is obtained by preserving a phase displacement of 120° between the gating pulses delivered to each leg of the inverter. In Fig. 3.1 the thyristors are numbered in their correct firing sequence. This means that they must be gated at uniform intervals in the sequence TH1, TH2, TH3, TH4, TH5, TH6 in order to complete one cycle of the output voltage. The output frequency of the inverter is determined by the gating frequency of the thyristors.

3.2.1. *Voltage Waveforms*

Taking the negative d.c. line as reference point N, and assuming instantaneous commutation, the terminal voltages v_{RN}, v_{SN} and v_{TN} have the square waveforms shown in Fig. 3.2. As in the single-phase circuit, v_{RN} equals the supply voltage V_d when TH1 conducts, and is zero when TH4 conducts. The feedback diodes again clamp the output voltage to the d.c. supply voltage and so maintain a flat-top square-wave output voltage on reactive loads.

The line-to-line voltages are obtained by subtraction. Thus, $v_{RS} = v_{RN} - v_{SN}$, $v_{ST} = v_{SN} - v_{TN}$ and $v_{TR} = v_{TN} - v_{RN}$. Since the line voltage is obtained as the difference of two square waves it contains no third harmonic or multiples thereof, because these are zero-sequence harmonics and, therefore, are in phase in each of the three square-wave voltages, v_{RN}, v_{SN} and v_{TN}.

The line voltage waveforms are shown in Fig. 3.2 and are described as "stepped waves" or "quasi-square waves". Fourier analysis gives the expression

$$v_{RS} = \frac{2\sqrt{3}}{\pi} V_d \left\{ \sin \omega t - \frac{1}{5} \sin 5\omega t - \frac{1}{7} \sin 7\omega t + \frac{1}{11} \sin 11\omega t + \cdots \right\}. \tag{3.1}$$

The r.m.s. value is $\sqrt{(2/3)}\, V_d$ or $0{\cdot}816 V_d$ and the fundamental component has an r.m.s. value of $\sqrt{6} V_d/\pi$ or $0{\cdot}78 V_d$.

In the case of a star or wye-connected load, the line-to-neutral voltage has the waveform shown in Fig. 3.3 with steps of $V_d/3$ and $2V_d/3$. The harmonic content of the line and phase voltages is the same, and the difference in waveshape is due to a different phase relationship between the fundamental and the harmonics. The complete expression for the phase voltage is

$$v_{RO} = \frac{3}{\pi} V_d \left\{ \sin \omega t + \frac{1}{5} \sin 5\omega t + \frac{1}{7} \sin 7\omega t + \frac{1}{11} \sin 11\omega t + \cdots \right\}. \tag{3.2}$$

These line and phase voltage waveforms are described as six-step waves since they are generated by an inverter with six uniformly spaced commutations per cycle. The lowest

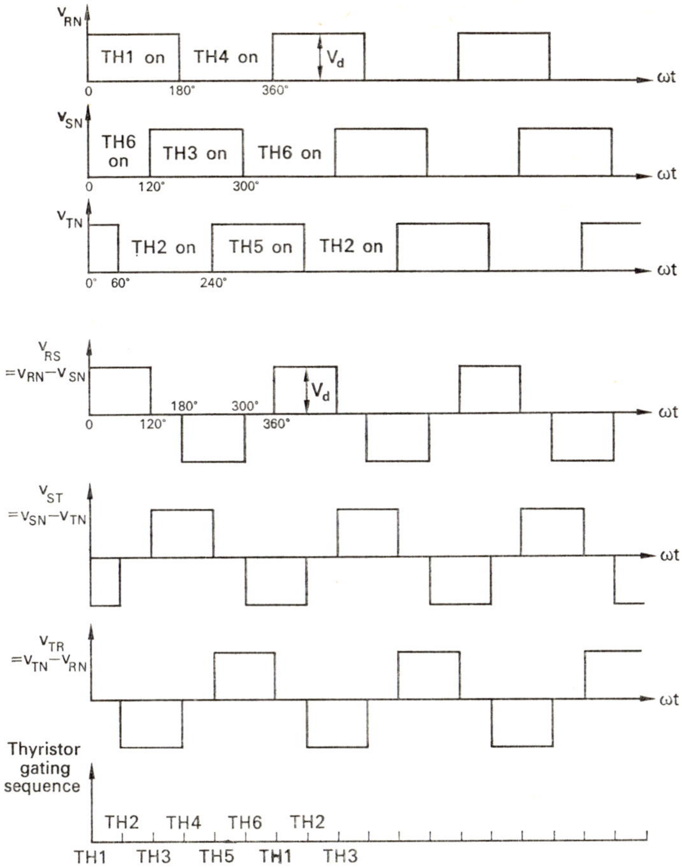

FIG. 3.2. Voltage waveforms and gating sequence for six-step operation of the three-phase bridge inverter.

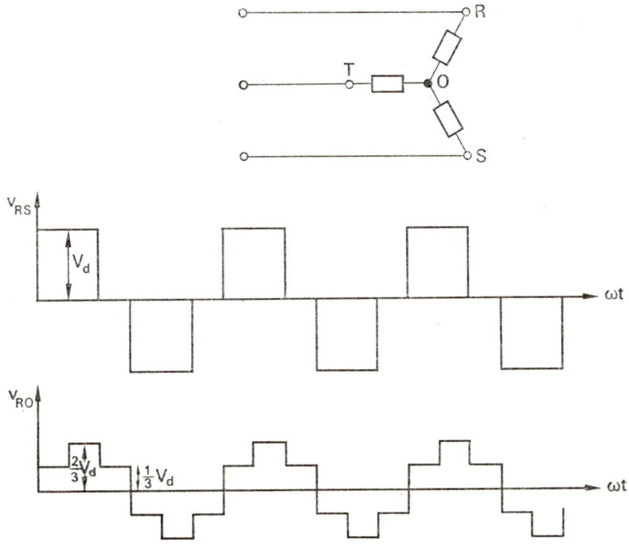

FIG. 3.3. Six-step line-to-line and line-to-neutral voltages for a star-connected load.

harmonic present is the fifth harmonic with an amplitude equal to 20 per cent of the fundamental. In practice, the six-step waveform has been used successfully in many adjustable-speed a.c. drives. The elimination of the zero-sequence harmonics is advantageous as these can be seriously detrimental to motor performance. The actual voltage waveform may differ somewhat from the ideal waveshape due to commutation effects and internal voltage drops in the inverter circuit.

3.2.2. *Current Waveforms*

The various current waveforms in the inverter circuit will now be investigated assuming a balanced delta-connected inductive load and instantaneous commutation. The voltage waveforms of Fig. 3.2 emphasize the fact that the six-step inverter is essentially a switching arrangement in which each phase of the load has an applied voltage of $+V_d$, zero, or $-V_d$. The phase current, therefore, consists of a series of exponential changes in current produced by the step changes in applied voltage, and each exponentially growing or decaying current has the time constant of the inductive load. The positive directions of the line currents i_R, i_S, i_T and the phase currents i_A, i_B, i_C are indicated in Fig. 3.1.

Assume thyristors TH4, TH5 and TH6 have been conducting, and take a time origin at the instant when TH1 is gated and TH4 is turned off by the commutating circuit. The voltage applied across phase C of the delta-connected load then has the stepped waveform of Fig. 3.4(a) with the voltage $+V_d$ applied for the first 120° of the cycle. This produces the exponentially rising current denoted by i_1 in Fig. 3.4(b). If the load time constant is appreciable, the phase current is still increasing after one-third of a cycle when TH3 is gated and TH6 is turned off. This connects both terminals of phase C to the positive supply line and reduces the phase voltage to zero. However, the inductive current is able to circulate around the circuit composed of TH1 and D3. Thus, the phase current decays exponentially towards zero and is denoted by i_2 in Fig. 3.4(b). One half-cycle is completed after 180° when TH4 is triggered and the voltage V_d is applied across phase C in the reverse direction. The other two phases have similar current waveforms displaced by 120° and 240° respectively. The line current is obtained in the usual manner as the difference of two phase currents. Thus, $i_R = (i_C - i_A)$ and is shown in Fig. 3.4(e). Mathematical expressions for the various currents are readily determined by writing the circuit equation for each conducting interval and matching the boundary conditions.[64]

The line current of a star-connected load is similar to that shown in Fig. 3.4(e), since every star-connected load has a delta equivalent. A comparison between the phase voltage in Fig. 3.3 and the phase or line current in Fig. 3.4(e) emphasizes that the step changes in phase voltage produce corresponding exponential changes in current.

Thyristor and diode currents. In Fig. 3.4, a time origin has been chosen at the commutation instant when TH4 is turned off, its gate signal is removed and TH1 is gated on. However, the line current $i_R = (i_C - i_A)$ is still negative, and since TH4 is blocked, the line current during this period must be carried by feedback diode D1, thereby reducing the net current drawn from the d.c. supply. Thus, if $(i_C - i_A)$ is negative at the instant of commuta-

tion, the current i_A circulates through D1 and TH5 until i_C increases sufficiently to make $(i_C - i_A)$ positive. Thyristor TH1 then starts to conduct, provided a gate signal is applied. Consequently, on inductive loads the thyristor must be re-gated when the line current reverses or, alternatively, the gating pulse must be applied for the full 180° during which conduction may occur.

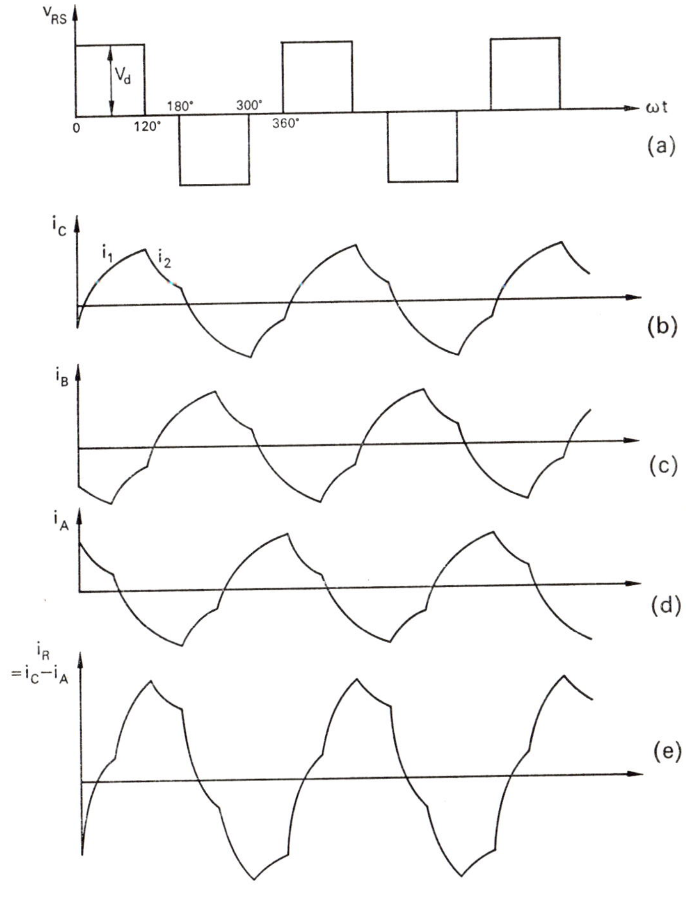

FIG. 3.4. Voltage and current waveforms in a three-phase, six-step inverter with a delta-connected inductive load: (a) line voltage; (b), (c), (d) phase, currents; (e) a.c. line current.

After a half-cycle, TH1 is turned off, but the line current remains positive for a time and must be carried by diode D4. Thus, on inductive loads the thyristor does not conduct immediately when the gating pulse is applied, but conduction must cease 180° later when the complementary thyristor is triggered. The thyristor and diode currents consequently have the waveforms shown in Fig. 3.5. The magnitude of the current flowing in thyristor TH1 at the end of its conduction interval is important, since this is the current which must be commutated. As the waveform is drawn in Fig. 3.5(b), this current is less than the peak thyristor current, but for inductive loads with low power factors the maximum thyristor current occurs at the instant of commutation.

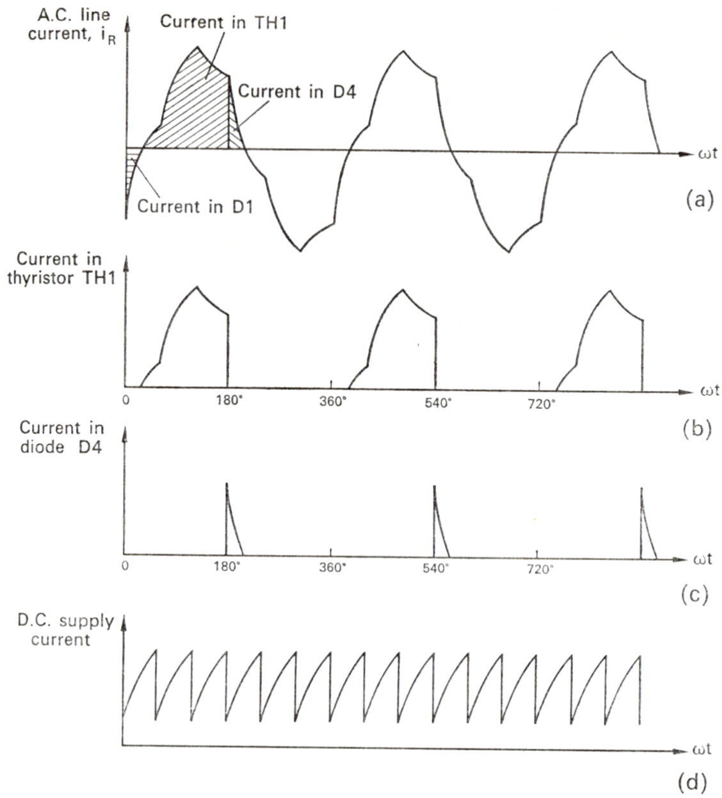

FIG. 3.5. Current waveforms in a three-phase, six-step inverter with an inductive load: (a) a.c. line current; (b) thyristor current; (c) diode current; (d) d.c. supply current.

D.C. supply current. The waveforms of the currents in phases A, B and C of the delta-connected load are shown in Fig. 3.4. In each phase, current component i_1 flows through the supply but the component i_2 does not, as it is simply the natural decay of the phase current circulating through a conducting thyristor and diode. The current delivered by the supply is, therefore, the sum of the i_1 components of the phase currents as in Fig. 3.5(d).

As explained above, the line current $i_R = (i_C - i_A)$ may be negative for part of the interval from zero to 60°, thereby indicating that current is flowing through feedback diode D1 and reducing the current drawn from the d.c. supply. In the interval from 60° to 120°, TH1 is the only gated thyristor connected to the positive d.c. line and, with very inductive loads, i_R may still be negative during this interval. Under these conditions, the d.c. supply current must reverse, indicating the return of reactive energy to the supply through the feedback diodes. In practice, this occurs when the power factor of the inductive load is less than 0·55. If the inverter obtains its supply from a rectifier, the d.c. current cannot reverse, and a capacitor must be connected across the d.c. link in order to absorb the reactive energy. A large capacitor is necessary to prevent an excessive rise in the inverter supply voltage.

3.3. HARMONIC NEUTRALIZATION

In a.c. motor applications, the static inverter delivers an output voltage of variable frequency, and this makes it extremely difficult to provide effective filtering. Since harmonic effects may prove objectionable, it is important to generate a waveform with a low harmonic content. Even in fixed-frequency applications where filtering is practicable, it is important to employ harmonic neutralization in order to reduce the size and cost of the energy-storage filter required.

3.3.1. *Stepped Waveforms*

In general, the output voltage waveform of a stepped-wave inverter consists of a series of steps, and each change in voltage level corresponds to a switching action in the inverter circuit. In the six-step waveforms of Fig. 3.3, the third harmonic, and multiples thereof, are eliminated, and the total harmonic content is 31·08 per cent of the fundamental, as compared with 48.4 per cent for a square wave. By using more complex circuits with larger numbers of thyristors, the number of steps in the voltage waveform can be increased to give a closer approximation to the ideal sinusoidal waveshape.

In general, assume a sine wave is being simulated by a stepped wave composed of N steps per cycle, each step occupying $2\pi/N$ radians. If the stepped wave is correctly formed, the only harmonics present in the waveform are of order $k = nN \pm 1$ where $n = 1, 2, 3, 4, \ldots$. Thus, the principal harmonics present in a six-step wave are the fifth and the seventh. In a twelve-step wave there are no harmonics lower than the eleventh, and an eighteen-step wave eliminates all harmonics up to the seventeenth.

In order to minimize the total harmonic content of the stepped wave, the voltage level of each step should be equal to the average value in a step interval of the sine wave being simulated.[65] Thus, the height of each step is proportional to

$$v_m = \int_{\theta_{m-1}}^{\theta_m} \sin\theta \, d\theta \qquad (3.3)$$

where

$$\theta_{m-1} = (m-1)\frac{2\pi}{N}, \qquad \theta_m = m\frac{2\pi}{N}$$

and

$$1 < m < N.$$

This results in a stepped waveform in which the lowest harmonic is one less than the number of steps per cycle, and the amplitude of the kth harmonic is $1/k$th of the fundamental amplitude. When these stepped waveforms are used in a three-phase system, a delta connection is possible, since the sum of the instantaneous voltages is always zero, as in the case of pure sine waves.

A twelve-step wave has the appearance shown in Fig. 3.6 and is commonly used in a.c. motor applications where the harmonic content of the six-step wave is excessive. It is readily shown by integration that the fundamental amplitude $V_1 = \pi V_m/3$, where V_m is the maximum step amplitude (Fig. 3.6). The other step amplitudes are $0{\cdot}268\,V_m$ and $0{\cdot}732\,V_m$. The Fourier series expansion of this twelve-step waveform is

$$v = \frac{\pi}{3} V_m \left\{ \sin \omega t + \frac{1}{11} \sin 11\omega t + \frac{1}{13} \sin 13\omega t + \frac{1}{23} \sin 23\omega t + \frac{1}{25} \sin 25\omega t + \cdots \right\}. \tag{3.4}$$

The total harmonic content is 15·22 per cent of the fundamental.

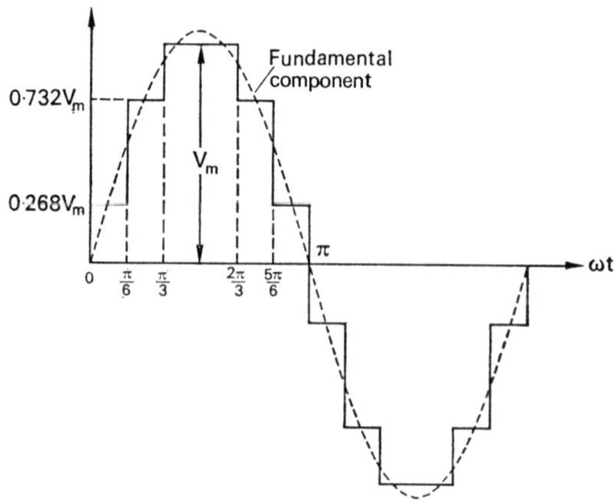

FIG. 3.6. The twelve-step voltage waveform.

Corresponding to the stepped wave defined by equation (3.3), there is also a second type of stepped wave with a zero voltage or dwell interval of one step width between $-\pi/N$ and $+\pi/N$.[66] The line-to-line and line-to-neutral voltages of Fig. 3.3 show the two types of six-step waveform, but a corresponding pair of waveforms exists for any value of N. The same harmonics are present in both waveshapes but their relative phase positions are different. This is evident from the Fourier expansions, as all harmonics in the first type of wave have positive signs, whereas certain harmonic terms in the second type have associated negative signs. Equations (3.1) and (3.2) for the two forms of six-step wave illustrate this point.

A stepped-voltage waveform is generated by adding a number of pulse-width modulated or square-wave voltages.[67] Each unit in Fig. 3.7 represents a single-phase bridge inverter of the type described in Section 2.3 and, as already explained, the pulse width of the inverter output can be adjusted by phase-shifting the gating of one side of the bridge relative to the

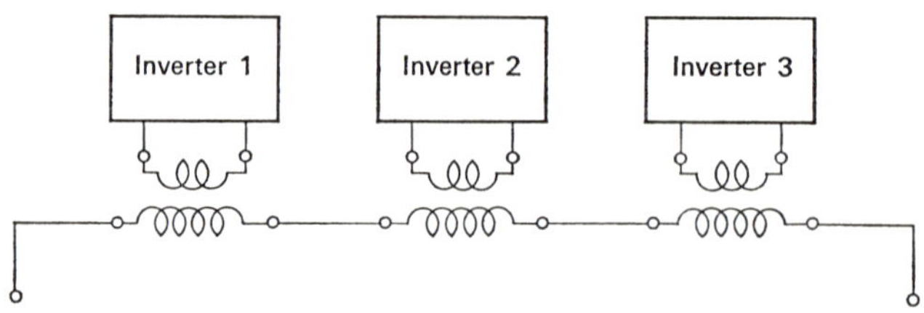

FIG. 3.7. The formation of a stepped-voltage waveform by series addition of inverter output voltages.

other. If the pulse-width of each inverter is suitably adjusted, and the output transformer ratio is correctly selected, the series addition of the secondary voltages gives the desired waveform. Alternatively, all inverters may operate in a square-wave mode with suitable phase-displacement between their outputs by means of the gating pulses.

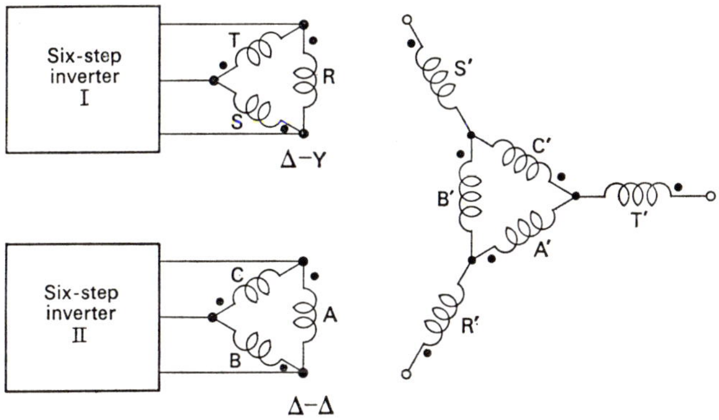

FIG. 3.8. The three-phase, twelve-step inverter circuit.

A three-phase, twelve-step waveform may be generated by the methods just outlined, but it can also be provided more economically by means of two six-step inverters which have a phase displacement of 30° and suitable output transformer connections (Fig. 3.8). Similarly, a three-phase, twenty-four-step output is generated using four six-step inverters which are phase displaced by 15°.[16] These complex schemes are uneconomical for low-power inverters but they are attractive at high power levels.

3.3.2. *Alternative Methods of Harmonic Cancellation*

Low-order harmonics may be eliminated from the inverter output by reversing the load voltage for selected intervals of each half-cycle.[68] Pulse-width modulation at a high switching frequency is also advantageous for harmonic reduction. When using this technique, the output voltage is rapidly switched on and off many times during a half-cycle. Pulse-width modulation is also used to vary the output voltage of the inverter, and is discussed under the heading of voltage control in Section 3.6 of this chapter.

3.4. FORCED-COMMUTATION METHODS

The development of simple and efficient methods of obtaining forced commutation has been the main problem in the field of static inversion. In earlier inverter circuits using thyratrons, commutation was achieved by means of large capacitors in parallel with the load.[69-70] Feedback diodes were not used, and this caused the output voltage waveform to vary considerably, depending on the frequency of operation and the power factor of the load. When the thyristor was developed, it was used initially to replace the thyratron in these parallel inverters. The thyristor turn-off time was about one-tenth of that for the thyratron and this permitted a significant reduction in the size of the commutating capacitor and gave a new impetus to the development of static inverter circuits.

Figure 3.9 shows a single-phase bridge inverter with parallel capacitor-commutation. Assume TH1 and TH4 are conducting and, as a result, capacitor *C* is charged with the left-hand plate positive. When TH2 and TH3 are gated, the capacitor discharges around two parallel paths, reducing the currents in TH1 and TH4 to zero and reverse-biasing them. The capacitor then charges with the reverse polarity through thyristors TH2 and TH3 and the series inductor *L*. A suitable choice of circuit values will ensure that sufficient recovery time is provided for the commutating thyristors. A similar sequence of events occurs 180° later when TH1 and TH4 are triggered.

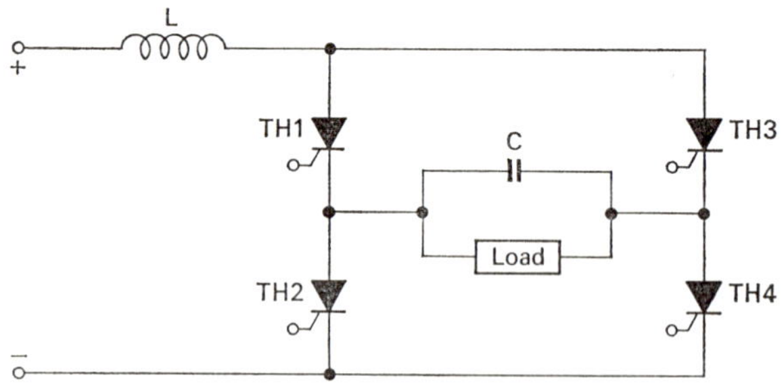

FIG. 3.9. The single-phase bridge inverter with parallel capacitor-commutation.

The absence of feedback diodes in this circuit means that the reactive energy associated with the load cannot be returned to the d.c. supply, but must be stored in the commutating capacitor and inductor. This demands excessively large circuit components and also causes the output voltage waveform to vary with load. The commutating voltage is a function of the load impedance, since the commutating capacitor is connected directly across the output terminals. Sufficient voltage may not be available for commutation under certain loading conditions and, in the case of highly inductive loads, additional commutating capacitance is necessary.

The simple parallel inverter is, therefore, unsatisfactory and does not fully exploit the advantages of the thyristor. Improved commutating circuits have been developed and all modern inverters use feedback diodes. The return of reactive energy to the supply permits reduced values of commutating inductance and capacitance and also eliminates variations in the output voltage waveform due to changes in load. In these inverters, the commutating circuit only affects the circuit operation when required, that is, during the commutation interval. An impulse of reverse current is produced to extinguish the conducting thyristor, and a reverse voltage is then applied until the thyristor has recovered its forward-blocking capability. This technique is known as impulse commutation and operates satisfactorily for wide variations in output frequency, power factor and load current. In an impulse-commutated circuit, the component values are determined by the peak current to be commutated, the available capacitor voltage, and the circuit turn-off time required.

3.4.1. *Impulse Commutation Methods*

In studying the operation of these circuits it will be assumed that the inverter is supplying an inductive load, because this imposes the most severe duty on the commutating circuit and also because this text is concerned with a.c. motor applications.

One leg of a bridge inverter is shown in Fig. 3.10 with the commutating components omitted. It is assumed initially that the upper thyristor TH1 is conducting and, therefore, terminal R of the load draws current from the positive terminal of the supply and returns it

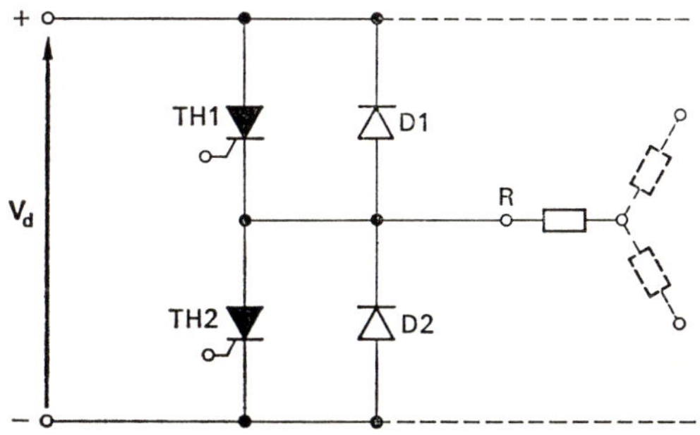

FIG. 3.10. One leg of a three-phase bridge inverter.

to the negative line through other conducting thyristors. When the commutation process is completed, thyristor TH1 is turned off, and the inductive load current continues to flow through feedback diode D2.

The following simplifying assumptions are also made:

1. the load inductance is sufficiently large to maintain the load current constant at a value I_o during the commutation interval. This is justified by the fact that the commutation process is completed in 10 to 100 microseconds;
2. the brief pulse of reverse thyristor current at turn-off can be ignored;
3. the turn-on time of the thyristor is negligible. This is a reasonable assumption since the turn-on time is only 1 to 4 microseconds, and the rate of increase of current is determined by external circuit components with much longer time constants;
4. the forward voltage drop of a conducting thyristor is negligible.

3.4.2. *McMurray–Bedford Circuit*[71]

Figure 3.11 shows one leg of a polyphase bridge inverter employing the McMurray–Bedford method of thyristor commutation. The centre-tapped inductor is small (of the order of 50 μH) and the two halves are assumed to be perfectly coupled. The commutation process is initiated when the complementary thyristor is triggered. Figure 3.12 shows the voltage and current waveforms for a lagging power factor load. The duration of the commutation interval B is exaggerated.

Interval A. Initially, thyristor TH1 is conducting and drawing load current from the positive side of the d.c. supply. In a highly inductive load, the rate of change of current is small, and the voltage across the commutating inductor, L_1, is negligible compared with the supply

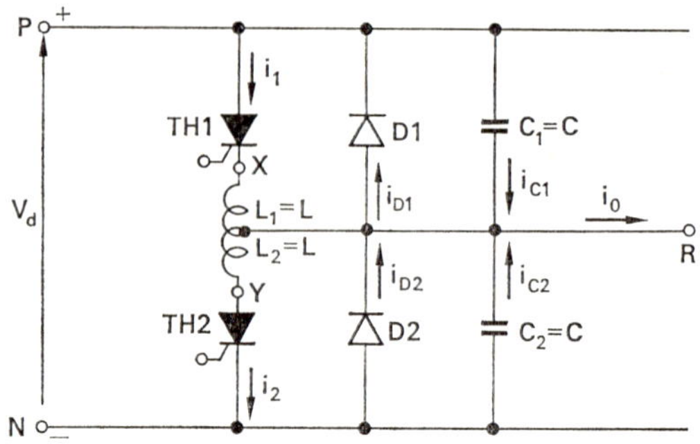

FIG. 3.11. One leg of a bridge inverter employing the McMurray–Bedford commutation method.

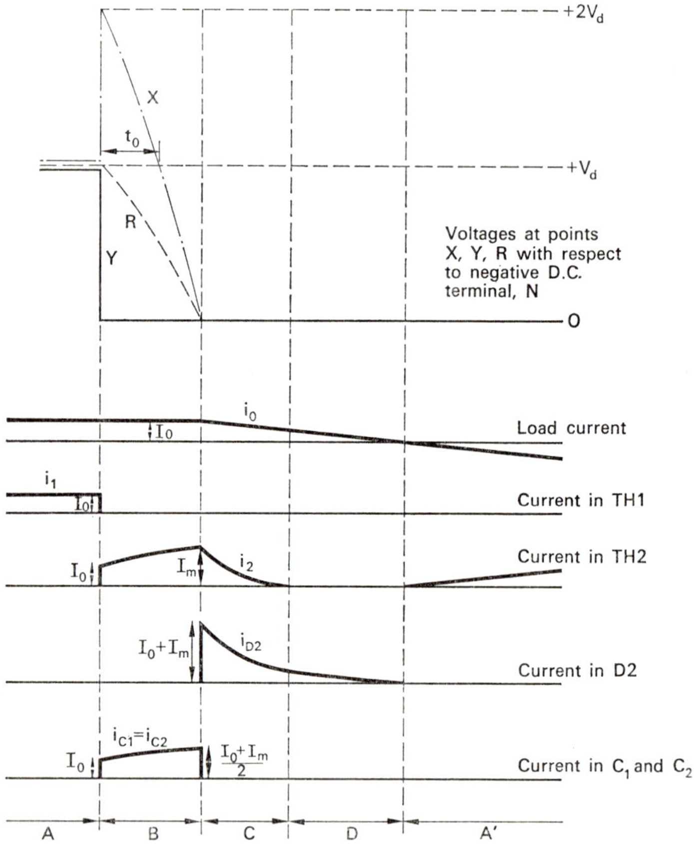

FIG. 3.12. Voltage and current waveforms in the McMurray–Bedford circuit with a lagging power factor load.

voltage. Terminal R of the load is, therefore, effectively connected to the positive supply line, and capacitor C_1 has zero voltage, while a similar capacitor C_2 is charged to the supply voltage V_d.

Interval B. When TH2 is triggered to initiate commutation, point Y is connected to the negative side of the supply. The voltages on capacitors C_1 and C_2 cannot change instantaneously, and hence a voltage V_d appears across winding L_2 of the commutating inductor. An equal voltage is induced in winding L_1, thus raising the potential of point X to $+2V_d$ with respect to the negative d.c. line (Fig. 3.12). This places a reverse bias of V_d on thyristor TH1, causing it to turn off. The load current I_o, which was flowing through TH1 and winding L_1, will transfer instantaneously to TH2 and winding L_2, in order to preserve the ampere-turns balance in the perfectly coupled transformer. The constant current I_o demanded by the inductive load is supplied by the charging currents of C_1 and C_2. These initial conditions determine the subsequent voltage and current variations during the commutation interval.

As capacitor C_2 discharges, the voltage applied across L_2 decreases, thereby reducing the voltage induced in L_1. Simultaneously, capacitor C_1 is charging up to the supply voltage and, after a short interval, the forward bias on TH1 due to C_1 exceeds the reverse bias due to L_1. If $L_1 = L_2 = L$ and $C_1 = C_2 = C$, a rigorous analysis shows that the voltage waveforms are portions of a sine wave of period $2\pi\sqrt{(2LC)}$, and the turn-off time, t_o, provided by the circuit is less than a quarter of a cycle.

During the commutation interval, the inductor L_2 supports the voltage on capacitor C_2. This causes the current i_2 in L_2 to increase from its initial value of I_o to a maximum value, I_m, when the voltage on C_2 has fallen to zero.

Interval C. Capacitor C_1 is now charged to the supply voltage, and the inductive load current i_o is drawn through feedback diode D2, thereby connecting load terminal R to the negative supply line. The inductor L_2 has stored or "trapped" energy which is dissipated in resistance losses by a current circulating in the closed circuit of L_2, TH2 and D2. Circuit design is usually optimized to reduce this trapped energy to a minimum, and extra resistance is sometimes added in series with the diode in order to speed the decay of circulating current and so permit a reduction in the current ratings of thyristor and diode.

Interval D. When the circulating current has decayed to zero, the commutation process is complete, but diode D2 continues to draw the inductive load current from the negative line.

Interval A'. The load current subsequently decays to zero and then reverses its direction. This blocks the rectifier D2, and the increasing line current in the new direction is carried by TH2, which must be re-gated if the original gating signal, which initiated commutation, has been removed. This completes the transfer of load current from TH1 to TH2.

The McMurray–Bedford circuit may be modified to permit the return of the trapped energy to the d.c. supply by means of a tapped output transformer. This improves the inverter efficiency, particularly at high operating frequencies when the number of commutations per second is large. However, the introduction of an output transformer increases the capital cost of the inverter.

3.4.3. *McMurray Circuit*

In the McMurray–Bedford circuit, the commutation process is initiated when the complementary thyristor is triggered. An alternative method due to McMurray[72] uses auxiliary thyristors to switch a high-Q inductance–capacitance circuit in parallel with the conducting thyristor. Figure 3.13(a) shows one leg of a bridge inverter using this principle, where TH 1′ is the auxiliary thyristor used to initiate the commutation of TH 1.

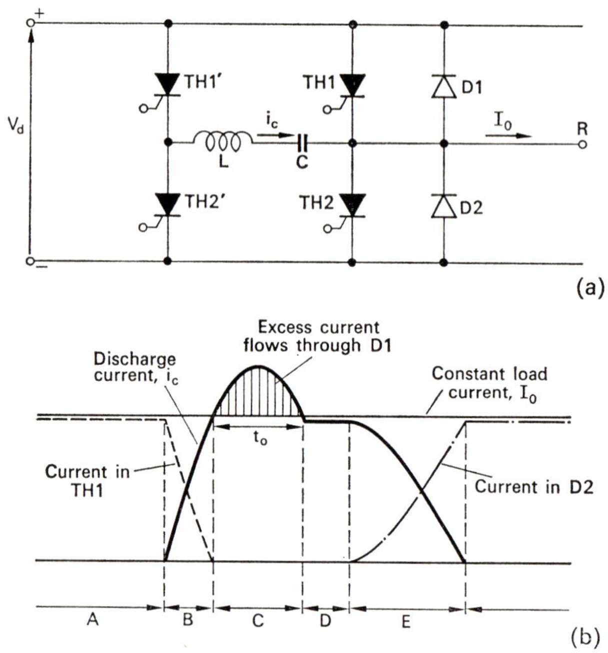

FIG. 3.13. One leg of a bridge inverter employing the McMurray commutation method: (a) basic circuit; (b) current waveforms with a lagging power factor load.

Interval A. Suppose TH 1 is conducting, thereby connecting terminal R of the load to the positive supply line. Capacitor *C* is already charged, with the right-hand plate positive, from the previous commutation of TH 2.

Interval B. When TH 1′ is triggered to initiate commutation, capacitor *C* discharges around the oscillatory circuit consisting of TH 1′, *L*, *C* and TH 1. As the discharge current i_c increases from zero on the first half-cycle, it gradually reduces the forward thyristor current to zero, as shown in Fig. 3.13(b).

Interval C. The load current I_o is maintained constant by the load inductance, and when the discharge current i_c exceeds I_o, the excess current flows through the feedback diode D1. Thus, thyristor TH1 has a reverse bias to turn it off, equal to the forward voltage drop of the feedback diode (about 1 V). The discharge current i_c rises to a peak value when the capacitor voltage is zero, and then decreases as the capacitor is charged in the reverse direction.

Interval D. Diode D1 ceases toconduct when i_c falls below the level of I_o and this removes the reverse bias from TH1. The load continues to draw a constant current through TH1′, L and C. This charges capacitor C with the reverse polarity until it is charged to the supply voltage V_d. Diode D2 then starts to conduct and clamps load terminal R to the negative supply line.

Interval E. The current I_o flowing in L continues to charge capacitor C until the trapped magnetic field energy in L is converted to electrostatic field energy in the overcharged capacitor. During this interval, D2 draws an increasing fraction of the load current, and the commutation transient is complete when all the load current flows through the diode.

As before, the load current subsequently decays to zero and reverses, thereby blocking diode D2 and bringing thyristor TH2 into conduction provided a gating signal is applied.

In the McMurray circuit, the turn-off time of the main thyristor must be shorter than the time, t_o, for which the discharge current i_c exceeds the load current, I_o. This circuit differs from other commutating circuits in the fact that the reverse voltage applied to the thyristor is limited to a volt or two, and the reverse current at turn-off is limited by the available discharge current. In the McMurray–Bedford circuit, the commutating voltage applied across the thyristor has approximately the same magnitude as the supply voltage, and the reverse current is limited by circuit inductance.

The no-load commutating losses in the McMurray circuit are small, and high inverter efficiencies can be achieved. The voltage on capacitor C at the end of the commutation transient increases with load current, I_o, and thus the commutating capability of the circuit automatically increases with load. The McMurray–Bedford circuit, on the other hand, has large commutating losses at no-load, resulting in a low inverter efficiency under light-load conditions. Another impulse-commutated inverter due to McMurray attempts to combine some of the better features of the two methods.(59)

3.4.4. *Input-circuit Commutation*

Instead of commutating each thyristor individually, it is also possible to commutate the inverter circuit as a whole.(20,73,74) When all the thyristors have been turned off, the next set of thyristors in sequence is gated. This method of commutation is called input-circuit commutation or d.c.-side commutation, and a typical circuit is shown in Fig. 3.14.

When thyristor THA is gated, the single commutating capacitor C is charged by an auxiliary d.c. supply with the upper plate negative. The inductor L_A produces an oscillatory overshoot so that the capacitor voltage exceeds the auxiliary supply voltage. When commutation is required, thyristor THC is gated, thereby connecting the negatively charged capacitor between the positive and negative rails of the thyristor bridge. The series inductors L_P

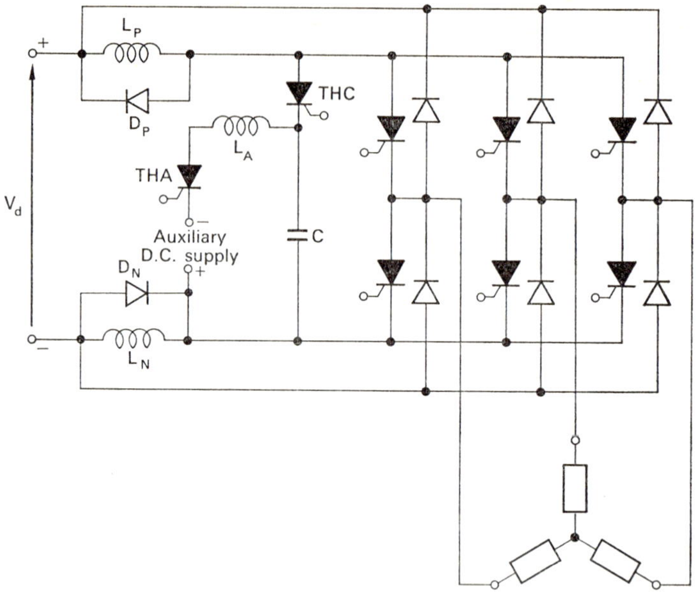

FIG. 3.14. Three-phase bridge inverter with input-circuit commutation.

and L_N are placed in the d.c. supply lines to drop the supply voltage. The feedback diodes are returned direct to the supply, causing the inductive load current to clamp the load voltage to the supply voltage, V_d. The negative voltage on capacitor C reverse biases the previously conducting thyristors for a sufficiently long interval to achieve turn-off. The capacitor charges up to the supply voltage, V_d, but does not overshoot, since the free-wheeling diodes D_P and D_N come into conduction and dissipate the excess energy stored in L_P and L_N. Thyristor THC now turns off and the bridge thyristors are re-gated in the next sequence. THA is also gated to charge the capacitor for the next commutation.

3.4.5. *Summary*

The operation of some typical impulse-commutated inverters has been examined in the preceding pages. Many other variations are possible, and considerable effort is being directed towards the development of more efficient and economic circuits.[75–77] As the cost of semiconductor devices continues to decrease, the tendency is to use more auxiliary thyristors and diodes in order to achieve a reduction in the number of magnetic components required. However, the basic principles of operation do not alter and, in each of the preceding examples, the commutating circuit must supply the current impulse required to turn-off the previously conducting thyristor and then apply a reverse bias until it has recovered its forward blocking ability.

Much of the energy stored in the capacitors prior to commutation is dissipated in the circuit during the commutation process.[77] Consequently, in a forced-commutated inverter, the commutating losses are usually appreciably greater than the conduction losses in the thyristors. As the operating frequency increases, the number of commutations per second increases proportionately, and the inverter efficiency falls. At power frequencies, inverter efficiencies of 85 to 95 per cent are readily achieved and reasonable efficiency is obtained up to frequencies approaching 1 kHz. Special inverter-type thyristors with short turn-off times are now available. These permit a reduction in the commutating capacitance, giving a more compact inverter with reduced commutating losses.

The turn-off time which is provided by the commutating circuit increases with the size of the commutating capacitor and the voltage to which it is charged, and it also varies inversely with the current being commutated. If the load current is excessive, sufficient recovery time is not available and a commutation failure occurs, resulting in a short-circuit condition. Once the inverter has been constructed with a certain commutating capacitance, the peak output current is limited by the commutating capability of the system. A certain amount of overcurrent capacity must be present to prevent a commutation failure on transient overloads, but excessive commutating capability should not be specified, as this reduces the inverter efficiency under normal operating conditions. Thus, the commutating capability of an inverter may be doubled by re-designing the commutating circuit, but the continuous rating is unaffected, and the commutating losses at the continuous rating are twice as large as before.

3.5. FREQUENCY CONTROL

The inverter output frequency is determined by the rate at which the thyristors are triggered into conduction. The pulse rate is set by a low-power reference oscillator which generates a train of timing pulses. These pulses are directed by logic circuits to the various thyristor gating circuits to control the generation of the firing pulses. The timing pulses are also used to control the turn-off circuits. The inverter output frequency is, therefore, determined solely by the reference oscillator and is unaffected by load transients. The frequency stability and accuracy requirements of the inverter dictate the type of reference oscillator required. A simple resistance–capacitance relaxation oscillator in a temperature-controlled environment has a stability of $\pm 0{\cdot}05$ per cent of any set frequency for periods of 48 hours. Frequency variation over a wide range is readily obtained by means of a single potentiometer in the oscillator circuit. When extreme frequency stability is specified, a crystal-controlled oscillator and digital count-down techniques may be used. Such systems can provide a stability of $\pm 0{\cdot}001$ per cent, or better.

The frequency control is an open-loop system and there is no feedback of information to the reference oscillator regarding the output frequency of the inverter. Consequently, the usual closed-loop stabilization problem is eliminated and the inverter has zero frequency regulation since steady-state or transient changes in load do not affect the oscillator frequency. Similarly, moderate changes in a.c. supply frequency and voltage do not alter the inverter output frequency. This accurate and stable frequency is one of the main advantages of static a.c. supplies. The precise frequency output is converted to a precise shaft speed when synchronous or reluctance motors are powered by the static inverter.

3.6. VOLTAGE CONTROL

When an a.c. motor has a variable-frequency supply, its terminal voltage must be varied as a function of the applied frequency in order to maintain constant magnetic flux density (Section 1.3). Consequently, the static inverter must deliver an adjustable alternating voltage whose magnitude is determined by the output frequency. This voltage variation can be achieved in one of the following ways:

(a) by variation of the alternating voltage output from the inverter;

(b) by variation of the direct voltage input to the inverter;

(c) by switching techniques within the inverter circuit.

Each of these methods has been used commercially and will now be considered.

3.6.1. *Control of Alternating Voltage Output from the Inverter*

Variable-ratio output transformer. This is one of the simplest methods of achieving voltage control. The inverter feeds the a.c. motor through a variable-ratio auto-transformer, and the transformer tapping is adjusted automatically in a simple closed-loop system (Fig. 3.15). A frequency-to-voltage converter produces a reference voltage which is proportional to the

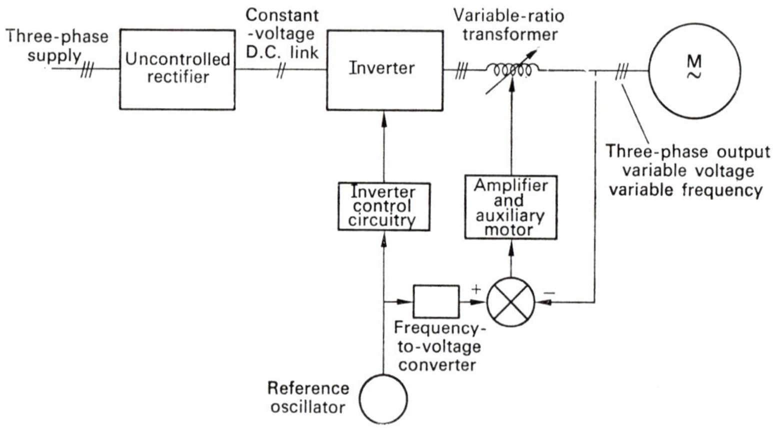

FIG. 3.15. Voltage control by means of a variable-ratio output transformer.

reference oscillator frequency. This signal is compared with a certain fraction of the output voltage. The difference is amplified and used to drive an auxiliary motor which adjusts the transformer tapping to reduce the difference. In practice, this gives a constant volts/Hz output within ± 1 per cent.

This system of voltage control has the advantage that the inverter output waveform does not vary over the frequency range. A high input power factor is achieved by using an uncontrolled diode rectifier input. The system is also advantageous when motors are run-up from standstill. By setting the transformer in the low-frequency, low-voltage position, the high inrush current of the motor is stepped down in the transformer primary. The starting current capability of the system is, therefore, considerably greater than the inverter current rating. On the other hand, the transformer has a constant primary voltage and a variable frequency. In order to avoid magnetic saturation it must be designed for the lowest operating frequency. This results in a rather bulky component when a large speed range is required, and consequently the lowest operating frequency is usually 10 or 15 Hz.

A further disadvantage of this scheme is the introduction of an electromechanical device into an otherwise static power supply. This causes a deterioration in the dynamic response of the system, since the adjustment time of the transformer is of the order of seconds. However, in many applications a fast transient response is not required, and reliable drives of this type have been marketed with full-load speed ranges of 4 to 1, or more. The cost of the auto-transformer and its voltage regulator is largely offset by the low cost of a simple inverter without any complexity in its power or control circuitry.

3.6.2. *Control of Direct Voltage Input to the Inverter*

The magnitude of the alternating voltage output from the static inverter is proportional to the direct voltage input from the rectifier. Voltage control can, therefore, be achieved by operating the converter with a variable d.c. link voltage. This system also maintains the same output voltage waveform over a wide frequency range, but commutation difficulties arise with some inverters, since the load current which can be commutated is proportional to the voltage on the commutating capacitors. In circuits which derive this voltage from the d.c. supply to the inverter, the commutating capability decreases when the input voltage is reduced. A constant-torque motor drive requires practically constant current at all speeds, and hence the inverter must commutate full-load current at low frequencies when the d.c. link voltage is reduced. If the commutating circuit is designed to operate satisfactorily under these conditions, the commutating losses are excessive at higher voltages and frequencies. For these reasons, static inverter drives of this type are limited to speed ranges of 3 or 4 to 1. For wide-range speed control, an auxiliary fixed-voltage d.c. supply is required to charge the commutating capacitors. This ensures that the commutating capability is independent of the direct voltage input to the inverter and, as a result, frequency ranges of 100 to 1 are obtained. The speed range at continuous rated torque may be less than this figure due to the motor speed regulation and the reduced ventilation at low speeds. Nevertheless, a full-load speed range of 20 to 1, or more, can be readily achieved.

The variable d.c. supply voltage can be obtained in a number of ways.

Variable-ratio input transformer. In this arrangement, a variable-ratio transformer is connected on the input side of the rectifier. Adjustment of the transformer tapping is obtained automatically in a closed-loop system similar to that used with an adjustable output transformer. However, in this case, the transformer operates at a.c. line frequency and is designed for this specific frequency, giving a more compact and economic unit.

Phase-controlled rectifier input. A static adjustable-voltage d.c. supply with a fast transient response is obtained by using a phase-controlled rectifier. At reduced output voltages, the input power factor is poor and the harmonic content of the output waveform is high. An output filter is necessary to smooth the direct voltage before feeding it to the inverter

(Fig. 3.16), but this also introduces a time lag into the system. Response times of the order of milliseconds are obtained in practice, and a half-controlled bridge rectifier is often used for economic reasons.

A closed-loop system of voltage control is again employed (Fig. 3.17). The reference voltage from the frequency-to-voltage converter is compared with a certain fraction of the d.c. link voltage and the difference is amplified and used to control the rectifier gating so that the required volts/Hz output is obtained.

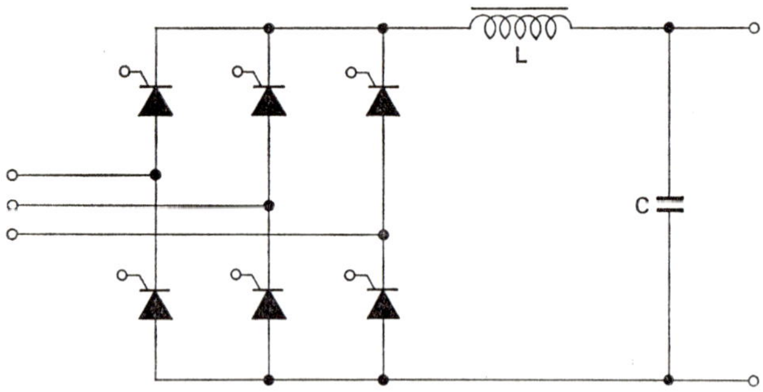

FIG. 3.16. Phase-controlled three-phase bridge rectifier and smoothing filter.

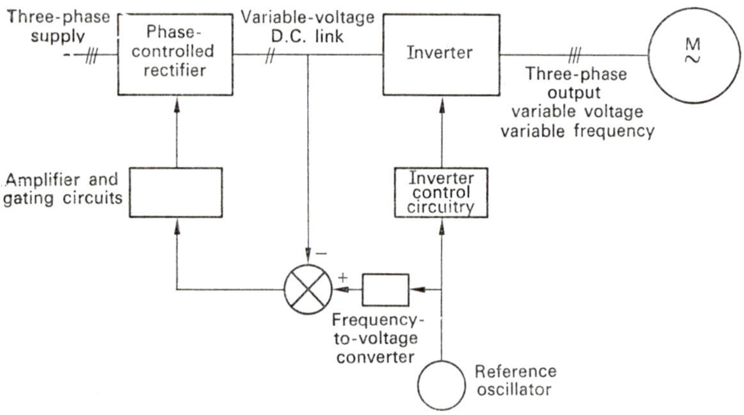

FIG. 3.17. Closed-loop control of d.c. link voltage.

Rectifier and d.c. chopper. A variable direct voltage input for the inverter may also be obtained by rectifying the three-phase a.c. supply using an uncontrolled diode bridge rectifier and then regulating the output by means of a d.c. chopper. In a series chopper circuit, a thyristor is placed in series with the d.c. supply and is rapidly switched on and off so that the output voltage consists of a series of rectangular pulses.[78] Figure 3.18(a) shows an a.c. motor speed control scheme in which the chopper thyristor, TH1, is alternately switched on by a gate pulse and then turned off by a commutating circuit of the type used in forced-commutated inverters. The idealized output voltage waveform of the chopper circuit is shown in Fig. 3.18(b). The average output voltage is usually varied (1) by keeping the total time per cycle, T, constant and varying the ratio of the on-time of the thyristor, T_1,

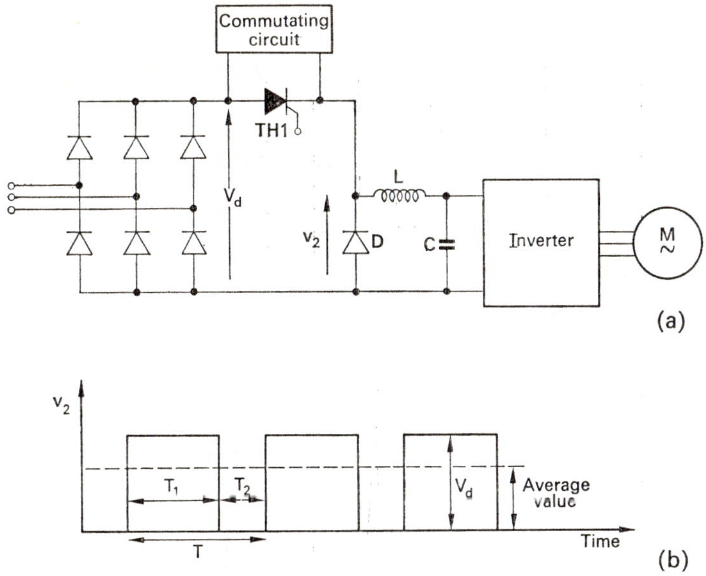

FIG. 3.18. Solid-state a.c. motor drive with voltage control by means of a chopper in the d.c. link: (a) basic circuit; (b) idealized output voltage waveform for the chopper circuit.

to the off-time, T_2, or (2) by holding the on-time, T_1, constant and varying the off-time, T_2. If the pulsating output voltage is filtered, a variable direct voltage is obtained, which can be used for speed control of a d.c. machine or as the input voltage to a static inverter for a.c. motor control.

Figure 3.19(a) shows a typical d.c. chopper circuit with a constant input voltage, V_d. Thyristor TH1 is the main series thyristor which is alternately switched on and off in order to vary the output voltage, v_2. Diode D is the usual free-wheeling diode to allow continuous output current, i_2, when TH1 is turned off. The series inductor, L, and shunt capacitor, C, filter the output voltage. An auxiliary thyristor, TH2, is triggered to initiate turn-off of TH1 and initially, when TH2 is gated, the capacitor C_1 charges up to the input voltage V_d through the load circuit. Thyristor TH1 is then gated to deliver current to the load, and C_1 discharges around the oscillatory circuit consisting of L_1, C_1 and diode D1. After the first half-cycle of this oscillation, the capacitor is charged to a voltage V_c with the negative plate at the anode of TH1. If the oscillatory circuit is lossless, V_c is equal to the input voltage, V_d. The reversal of current flow for the next half-cycle of the oscillation is prevented by diode D1 and, consequently, capacitor C_1 retains the voltage, V_c. At some subsequent time, when TH2 is gated, the capacitor C applies a reverse bias across TH1 and turns it off. The output current, i_2, is maintained by the filter inductance and flows through C_1 and TH2. This current gradually reduces the voltage on C_1 to zero and then charges it up with the opposite polarity. If C_1 is large enough, the forward voltage is not re-applied across TH1 until it has regained its forward-blocking capability. When C_1 has charged up to the supply voltage V_d, TH2 becomes reverse biased and turns off. The filter inductance maintains the output current i_2 by drawing the current through the free-wheeling diode D, but the output voltage v_2 is zero until TH1 is re-gated and the cycle of operations is repeated. Figure 3.19(b) shows

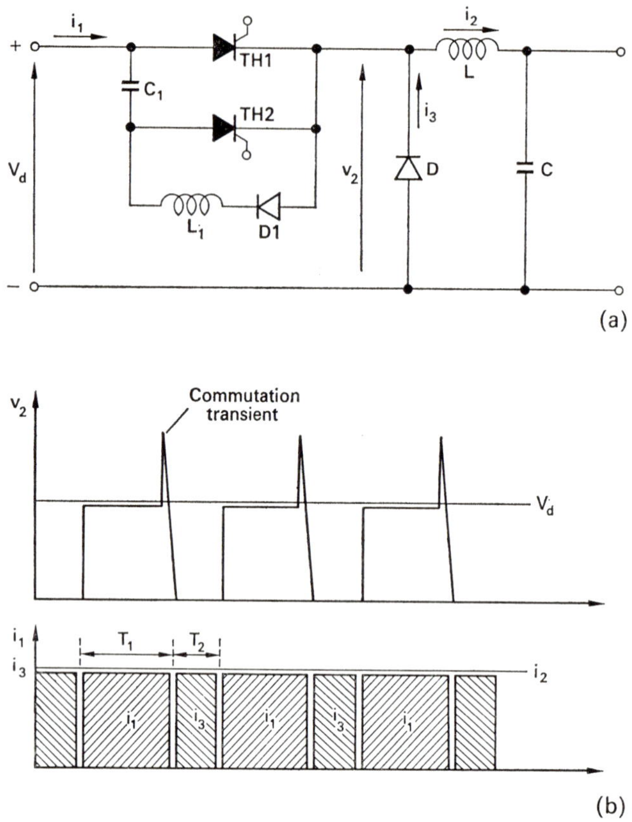

FIG. 3.19. Typical d.c. chopper for voltage control: (a) basic circuit; (b) voltage and current waveforms with a large filter inductance.

the voltage and current waveforms for a large filter inductance which maintains a constant output current.

The d.c. chopper is a static system and consequently has a high speed of response. By choosing a fast chopping frequency, the filter requirements are eased and the time lag in the system response is reduced. However, energy is lost at each commutation, and the efficiency decreases as the pulse repetition frequency is increased. In practice, chopping frequencies of a few hundred Hz to several kHz are used.

When a d.c. chopper is used to provide voltage control in a static a.c. drive, the volts/Hz output of the inverter is determined by a control voltage, which is derived from the reference oscillator and is fed to the gating and commutating circuits of the chopper. The closed-loop system shown in Fig. 3.17 for the phase-controlled rectifier is also used in this case.

3.6.3. *Voltage Control by Switching Techniques within the Inverter*

The fundamental component of the inverter output voltage can be varied by modifying the output waveform using switching techniques within the inverter circuit. The harmonic content usually increases as the output voltage is reduced, but the system is completely static and has a high speed of response.

Phase-shift voltage control. Phase-shift control requires two similar inverters which are operated at the same frequency from the same d.c. supply. The inverter outputs are combined in a transformer, and voltage control is obtained by phase-shifting the output of one inverter relative to the other. This is achieved statically by phase displacement of the gating pulses delivered to each inverter. The inverter output voltages may be added by means of two output transformers with their secondary windings connected in series. When the two sets of gating pulses are in phase, the fundamental output voltage is twice that of a single inverter. When the gating signals are displaced by an angle γ, the fundamental output voltage is reduced by a factor cos $(\gamma/2)$, and a phase displacement of 180° reduces the output voltage to zero. A signal from the reference oscillator controls the phase displacement so that the required output volts/Hz characteristic is obtained. The usual closed-loop system is employed.

Figure 3.20 shows the voltage waveforms obtained when the outputs of two six-step inverters are added.[79] With zero phase displacement, the phase-to-neutral voltage has a six-step waveshape and the fundamental component has its peak value. When phase-shift is introduced, the voltage waveform changes and the fundamental component is reduced.

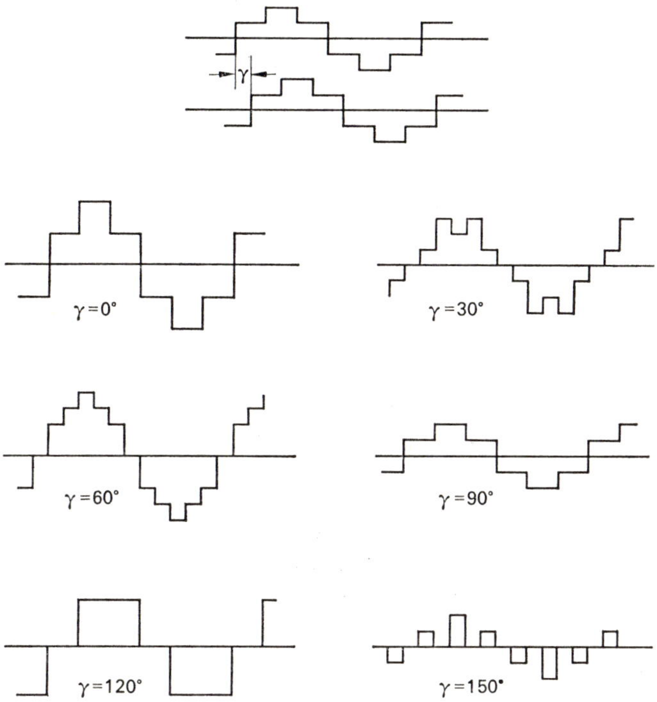

FIG. 3.20. The addition of two six-step voltage waveforms with an adjustable phase displacement, γ.

No harmonic which is eliminated from the individual inverter output waveforms is reconstituted when phase-shift control is applied. Thus, if two twelve-step voltage waveforms are combined in a phase-shift inverter, no harmonics of lower order than the eleventh will appear in the output voltage. However, the magnitudes of the unneutralized harmonics vary with the phase displacement γ and are relatively large compared with the fundamental for large values of γ.

Phase-shift control is obtained, in practice, using the circuit of Fig. 3.21, in which a three-phase output transformer has its primary phases connected between corresponding terminals of two six-step inverters. A six-step output voltage appears at terminals R, S, T and also at terminals R′, S′, T′. If the two waveforms are 180° out of phase, a six-step voltage

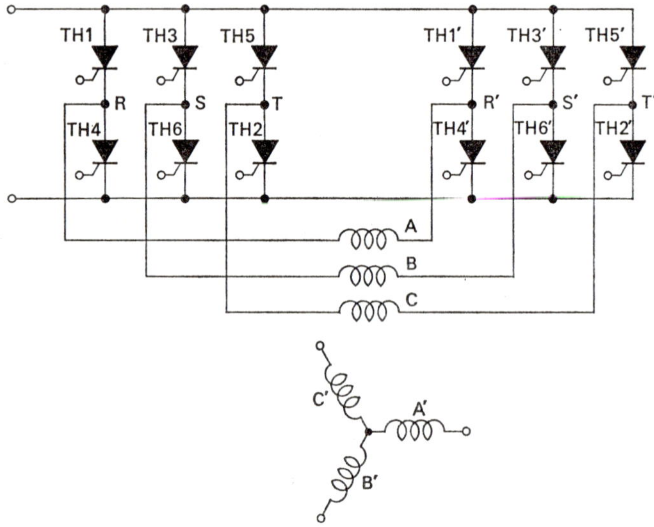

FIG. 3.21. Basic inverter circuit for phase-shift voltage control.

of maximum amplitude is applied across the primary windings. When the gating signals of one bridge are phase-shifted relative to the other, a reduced voltage is applied to the transformer primary.

This circuit may also be regarded as three single-phase bridge inverters, each supplying one primary phase of the output transformer. Thus, thyristors TH1, TH1′, TH4 and TH4′ constitute an inverter bridge supplying primary phase A. From this point of view, the phase displacement of the gating pulses of one side of the bridge relative to the other, results in the generation of a simple pulse-width modulated waveform, as explained in Section 2.3.1. A reduction in the pulse width of the primary voltage also reduces the magnitude of the fundamental component. The voltages delivered to phases B and C have the same waveform as phase A but are displaced by 120° and 240° respectively. Thus, phase-shift control and simple pulse-width control are equivalent, and similar output voltage waveforms are developed in each case.

In general, phase-shift methods of voltage control are rather complex and unattractive for low-power applications. However, both inverter circuits contribute to the power-handling capacity, and the complexity is justified in high-power applications when the required output power cannot be obtained using single thyristors. By using two independent inverters with a variable phase shift it is possible to incorporate voltage control at the expense of a more complex control circuit. The disadvantages of this method of voltage control are the high harmonic content and poor thyristor utilization at reduced output voltages. The distorted output voltage waveform may cause excessive heating in a.c. motors, and this sets a lower speed limit which is determined by the motor design. In practice, speed ranges of 5 to 1, or more, can be obtained.

Pulse-width modulation (PWM). The chopping or pulsing technique which is used to transform a direct voltage may also be used to control the alternating voltage output of a static inverter. The square or step-wave output voltage is rapidly switched on and off several times during each half-cycle so that a number of pulses of equal amplitude are formed. Each pulse has the amplitude of the inverter input voltage, V_d. Figure 3.22(a) shows an elementary form of multiple-pulse-width modulation in which the pulse width, T_1, and the off-time, T_2, are

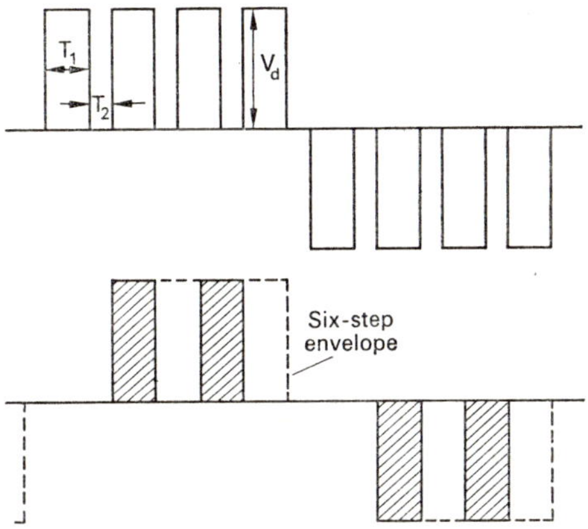

FIG. 3.22. Elementary forms of pulse-width modulation.

constant throughout the half-cycle. The magnitude of the fundamental output voltage is controlled by variation of the total on-time during a half-cycle. This may be achieved by maintaining a constant pulse width and varying the number of pulses per half-cycle, or alternatively, by variation of the pulse width for a constant number of pulses in each half-cycle.

The pulse-width modulation technique is readily applied to the single-phase bridge inverter in Fig. 2.4(a), since each side of the inverter can commutate independently, and an interval of zero voltage is obtained whenever the two load terminals are connected to the same d.c. line. By commutating one side of the bridge several times during a half-cycle, the output voltage waveform of Fig. 3.22(a) can be generated. A three-phase pulse-width modulated waveform is obtained using the original bridge inverter of Fig. 3.1.[80] The basic circuit is usually modified so that each thyristor has its own commutating circuit which can interrupt the flow of current at any desired instant. The commutating circuit of Fig. 3.19(a) can be used, but several alternative methods are available.[81] The pulse-width modulated waveform may also be generated by using input-circuit commutation to deliver a pulsating direct voltage to the inverter.

Elementary pulse-width modulated inverters are readily constructed to deliver a waveform with only two pulses per half-cycle within a six-step envelope, as in Fig. 3.22(b).

Appreciable low-order harmonics are introduced, principally the fifth and the seventh,[30] and these cause a significant deterioration in the low-speed performance of the a.c. motor. The harmonic content decreases as the number of pulses per half-cycle is increased and the waveform approximates more closely to the six-step wave. However, even with a very large number of pulses, the harmonic content of this type of waveform cannot decrease below the level associated with the six-step wave. In order to eliminate the low-order harmonics, more sophisticated techniques of pulse-width modulation are employed, in which high-frequency pulsing occurs throughout the half-cycle.

A.C. motor speed control usually requires a constant volts/Hz supply, that is, constant volt-seconds/cycle. Consequently, the area in volt-seconds enclosed by a half-cycle of the output voltage waveform must be the same at all frequencies. In many variable-speed drives this is achieved by maintaining a constant pulse duration as the fundamental output frequency is reduced (Fig. 3.23). The pulse repetition frequency therefore decreases linearly

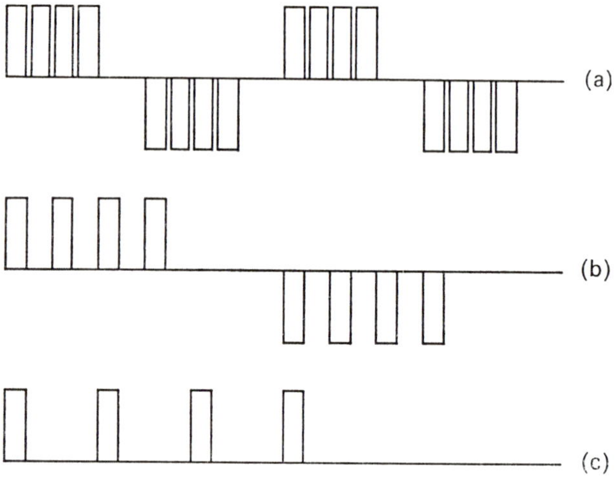

FIG. 3.23. The production of a constant volts/Hz output using a pulse-width modulated inverter. Output voltage waveform: (a) at maximum frequency; (b) at half-frequency; (c) at quarter-frequency.

with the fundamental output frequency and, if required, a voltage boost is obtained at low frequencies by increasing the pulse width. With this method of voltage control, the output voltage waveform deteriorates at low frequencies as the interval between pulses increases. The harmonic frequencies also decrease with the fundamental frequency, and the machine inductance is less effective in smoothing the current waveform. This sets a limit to the lowest speed of operation, but the machine performance can be improved by introducing external reactors in series with the stator leads. This causes some loss of fundamental torque, but the reactance presented to the harmonics is greater than that offered to the fundamental current, and hence the current distortion and harmonic losses are greatly reduced.

The severe voltage distortion at low output voltages can also be reduced by means of a step-down auto-transformer on the output side of the inverter. When the voltage distortion due to pulse-width modulation becomes excessive, the inverter voltage is restored to its full value and the step-down transformer is switched in. Pulse-width modulation is then applied, as before, to give reduced output voltages with an improved waveform. A series of transformer tappings is necessary to cover a wide range of voltages, and the transformer must be designed to operate with the full inverter voltage at a reduced frequency. By using a combination of series reactors and an auto-transformer, a.c. motor performance with a simple two-pulse or four-pulse waveform can usually be improved to almost equal that with a six-step waveform.[82] The step-down transformer also improves the starting performance by delivering large currents to the motor which can then develop considerable starting torque.

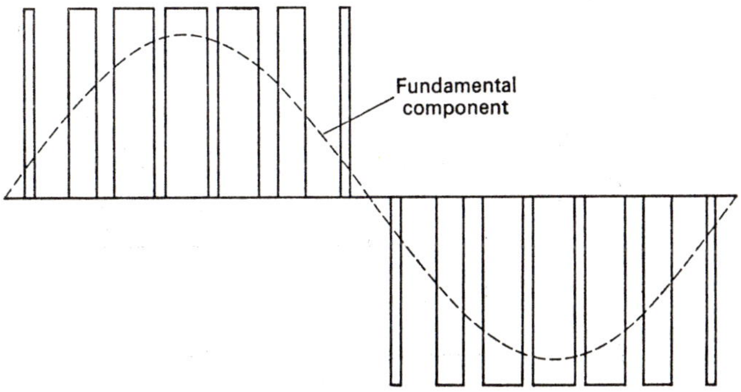

FIG. 3.24. Output voltage waveform of a pulse-width modulated inverter with sinusoidal modulation of the pulse width.

In more refined pulse-width modulation systems, the pulse width is varied throughout the half-cycle in a sinusoidal manner (Fig. 3.24). Strictly speaking, the pulses should be regularly spaced, and the pulse width at a particular position should be proportional to the area under the sine wave at that position. This is analogous to the generation of a stepped waveform in which the height of each step is proportional to the area under the sine wave in a particular interval (Section 3.3.1). In the pulse-width modulated waveform, the lowest harmonic frequency is at the pulse repetition frequency and, if this is much higher than the fundamental frequency, sufficient filtering is provided by the machine inductance. These waveforms are usually generated by means of a control circuit in which a high-frequency triangular waveform is mixed with a sinusoidal reference waveform of the desired frequency. Figure 3.25 shows the waveforms for a comparator circuit in which an asymmetrical triangular waveform is used to produce a suitable switching signal. This is fed to the thyristor gating circuitry and generates a pulse-width modulated waveform which approximates closely to the required waveshape.[30] A symmetrical triangular waveform has also been mixed with a sine-wave reference in order to develop the required control signal.[24] Voltage control is obtained by varying the widths of all pulses without upsetting the sinusoidal relationship. A control voltage, which is derived from the reference oscillator, determines the output volts/Hz in the usual closed-loop system.

When the pulse width is sinusoidally modulated, the output voltage and current have a small harmonic content. This reduces the undesirable torque pulsations which are usually associated with low-frequency operation of a static a.c. drive (Chapter 5). Mains interference and distortion are also minimized by the use of an uncontrolled diode bridge rectifier and a high pulse repetition frequency in the inverter. The pulse-width modulated system also requires a minimum of auxiliary equipment, since the only apparatus required, in addition to the inverter, is an inexpensive diode bridge rectifier. An output transformer is not essential unless several inverter outputs are combined for harmonic cancellation. This reduction in auxiliary equipment is obtained at the expense of a more complex inverter control circuit, but a full-load speed range of 20 to 1 is readily provided, and the dynamic response is excellent.

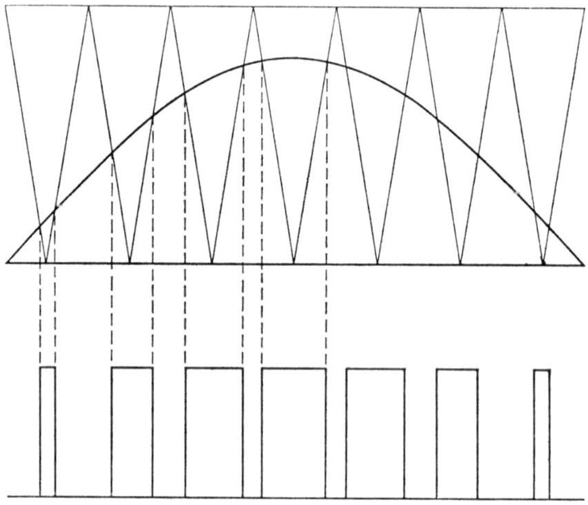

FIG. 3.25. Control circuit waveforms for a pulse-width modulated inverter.

In the pulse-width modulated inverter, as in the d.c. chopper, the choice of pulse repetition frequency is a compromise. A high pulsing frequency reduces the lower order harmonic content and eases the filter requirements. On the other hand, the inverter commutating losses are determined by the number of commutations per second, and this increases linearly with the pulsing frequency. In practice, when determining the pulse repetition frequency, the increase in inverter commutating losses must be balanced against the reduction in a.c. motor losses due to the reduced harmonic content of the stator voltage.

3.6.4. *Summary*

Several different techniques for inverter voltage control have now been described. Those systems which employ a fixed-voltage d.c. link have several advantages.

1. As already pointed out, many forced-commutated inverter circuits have commutating capacitors which are charged by the inverter supply voltage. In such circuits, a reduction in the d.c. link voltage causes a corresponding reduction in the commutating capability of the system. With a constant link voltage, rated current can be commutated even at very low output voltages (and frequencies) and hence the motor can develop rated torque at all speeds.

2. A constant link voltage can be provided by an uncontrolled diode bridge rectifier, and the absence of firing delay means that a high input power factor is presented to the a.c.

supply. The diode bridge rectifier is also the most simple and economic circuit available. It requires no auxiliary control circuitry and is extremely reliable.

3. The presence of a constant-voltage d.c. link permits the parallel operation of several independent inverters on the same rectified supply, with a considerable saving in rectifier costs. In addition, a stand-by battery or motor-generator set can be switched-in for emergency operation during a.c. supply failures. This system is commonly used for multi-motor drives in process lines where power supply interruptions cannot be tolerated (Chapter 7).

The advantages associated with a constant d.c. link voltage are present when phase-shift voltage control or pulse-width modulation is used. The rectifier and series chopper may also be regarded as a constant-link-voltage system if the chopper is considered as part of the inverter circuit, and the commutating supply is taken from the input side of the chopper. These static voltage control systems have great accuracy and flexibility and a high speed of response.

The use of a variable d.c. link voltage eliminates the advantages listed above, but it does provide a system in which the output waveform remains the same over the frequency range. In particular, if the commutating voltage supply is provided by an independent uncontrolled rectifier, the first advantage listed above is retained, and wide-range speed control is possible without any serious deterioration in low-speed peformance. The use of a variable-ratio output transformer retains the advantages of a constant link voltage but introduces a large time lag in the system response. However, this is not a serious objection in many applications, and the method of voltage control which is adopted in practice is determined by economic considerations and the system performance required.

3.7. CONTROL CIRCUITRY

As described in Section 3.5, the frequency of the a.c. output from a static inverter is controlled by the frequency of the gating pulses. A low-power resistance–capacitance oscillator is normally used as the reference or master oscillator, and the reference frequency is fed to the logic or control circuits. These generate suitable pulses and distribute them in the correct sequence to the thyristor gating and commutating circuits. In a bridge inverter, the thyristor gate electrodes have different d.c. potentials and the gating pulses must be supplied by transformers in order to provide d.c. isolation. Positive gate signals should not be applied while the anode voltage is negative, as this causes excessive junction heating and may result in thermal runaway.

In a stepped-wave inverter with inductive loading, the current does not transfer to the incoming thyristor immediately after the complementary thyristor has been commutated (Section 3.4). Instead, the load current continues to flow for a period through a feedback diode, and consequently the incoming thyristor must be re-triggered when the current reverses. The control circuit would become very complex if required to provide re-triggering at the correct instant for a variable load, and the usual solution is to apply a gate pulse for the complete half-cycle during which the thyristor may conduct. This presents problems when very-low-frequency operation is required, since the transmission of fundamental-frequency gating pulses would require very large gating transformers. The difficulty is overcome by using a high-frequency square-wave carrier signal (10–20 kHz) which is readily transmitted by a small transformer.[83] The input to the transformer is gated on and off by a low-frequency square-wave from the logic circuitry (Fig. 3.26). The carrier-frequency output signal at the transformer secondary is then full-wave rectified and filtered. The resulting low-frequency square wave is used to gate the thyristor.

A typical control circuit for a six-step inverter consists of a three-stage ring counter triggered by a reference oscillator of the resistance–capacitance relaxation type. The ring counter produces a six-phase square-wave output, and each phase is connected to one of

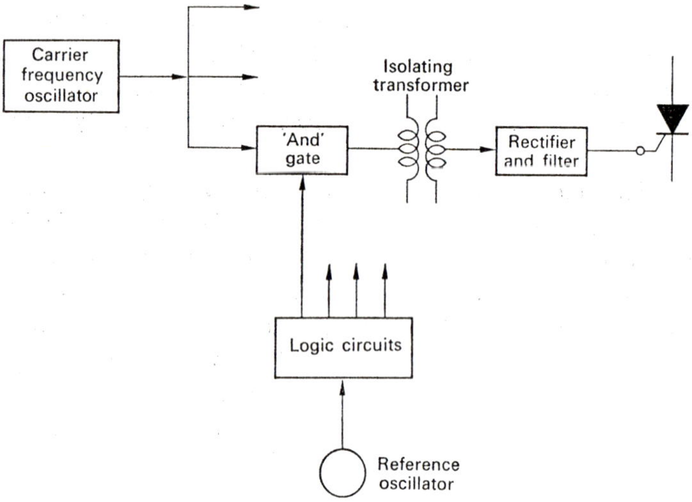

FIG. 3.26. Block diagram of the inverter gating circuitry.

the six thyristor gating circuits. In each of these circuits, the high-frequency carrier signal is "and"-gated with the square wave from the ring counter, and the output is delivered by the gating transformer and full-wave rectifier, to the gate of the appropriate thyristor. Inverters using harmonic cancellation or pulse-width modulation techniques require more complex control circuitry, but the introduction of integrated circuits is leading to increased reliability and reduced cost.

As explained in Section 3.5, very precise speed control can be obtained using a reluctance motor powered by a d.c. link converter. A precise output frequency is obtained by using a high-stability reference oscillator to determine the inverter frequency. The frequency control is a simple open-loop system, but the output voltage of the inverter is determined by a closed-loop voltage regulator. In order to obtain the required volts/Hz output from the inverter, an independent reference voltage from a potentiometer is fed to the variable-frequency reference oscillator and also to the voltage control circuit. Alternatively, the reference voltage is fed to the frequency control circuit only, and the output voltage is determined by the inverter frequency.

In an induction motor drive, very precise speeds are not expected, and independent control of the inverter frequency is not essential. In such applications, the reference voltage may be fed to the voltage control circuit, and the output frequency is then determined by the d.c. link voltage or the motor voltage. This type of control scheme has certain advantages.

Since the inverter frequency follows the voltage, a very rapid decrease in stator frequency is impossible, and consequently regeneration is limited, and the d.c. link voltage cannot rise excessively (see Section 3.8). This type of control system also minimizes the stability problems associated with inverter-fed induction motor drives(35) (Chapter 5). Closed-loop control of rotor slip frequency also provides a variable-speed induction motor drive with many desirable characteristics, and this is considered further in Chapter 6.

The control circuitry of a pulse-width modulated inverter can be modified to make the inverter behave as a constant-current supply.(19) The inverter thyristors are rapidly switched on and off so that the output current oscillates about the prescribed value. In this manner, the inverter can supply the motor with a sinusoidal or six-step current waveform of controlled amplitude. The advantages of this form of operation are also discussed in Chapter 6.

3.8. REGENERATIVE AND DYNAMIC BRAKING

The d.c. supply for the static inverter is obtained by rectification of the alternating supply voltage. The advantages of the simple diode bridge rectifier have been listed in Section 3.6.4 but the choice of rectifier circuit for a particular application depends on the method of voltage control used, and whether regenerative or dynamic braking is incorporated in the system.

If the variable-speed motor is overhauled by the load, regeneration occurs and power is returned to the d.c. link through the feedback diodes of the static inverter. This arises when the inverter output frequency is suddenly reduced, and also occurs in crane and hoist applications when lowering a load. The simple diode bridge circuit and the half-controlled bridge circuit are both incapable of inverter operation, and hence regenerated energy cannot be returned to the a.c. supply. However, if the rectifier supplies several inverters simultaneously, regenerated power which is returned to the d.c. link by one inverter will help supply the rectifier load. Phase-controlled inverter operation is only possible in a fully controlled bridge circuit of six thyristors. This circuit functions as a rectifier when the firing is delayed by less than 90°, and inverter operation is obtained when the firing is delayed by more than 90°. As explained in Section 2.5.3, the current direction cannot be changed, and the applied voltage polarity must be reversed.

When the d.c. link supply is obtained from a phase-controlled rectifier or a d.c. chopper, the output voltage waveform is usually filtered as shown in Fig. 3.16 and Fig. 3.18. The large capacitor across the inverter input terminals smooths the direct voltage supply for the inverter, and reduces the source impedance. When the inverter load operates regeneratively, power is returned through the feedback diodes, charging up the smoothing capacitor and raising the d.c. link voltage. In order to return this regenerated power to the supply network, the rectifier connections should be reversed and the firing angle increased beyond 90°. The switching of power leads is undesirable and is avoided by connecting another thyristor bridge in anti-parallel, as shown in Fig. 3.27. The second thyristor bridge is blocked during normal operation, but when regeneration occurs, it accepts the reverse current from the d.c. link circuit. Energy is then returned to the a.c. network if the delay angle is suitably adjusted. The anti-parallel or double-converter connection thus permits direct current and power flow in either direction, but it requires twice as many thyristors as the single-converter circuit, and the control circuitry is also more complex. The double-converter circuit is often used for speed control and braking of reversible d.c. motor drives.

The provision of regenerative braking obviously adds to the cost and complexity of the d.c. link converter. In many applications, the transient performance and operating economy offered by regenerative braking are not essential, and dynamic braking gives a satisfactory performance for less capital expenditure. In the dynamic method, the regenerated power is dissipated in an external resistor. This is switched across the d.c. link supply when the link voltage reaches a certain level due to regenerative charging of the filter capacitor. The switching may be achieved automatically by means of an auxiliary thyristor which is triggered from a voltage-sensing circuit.

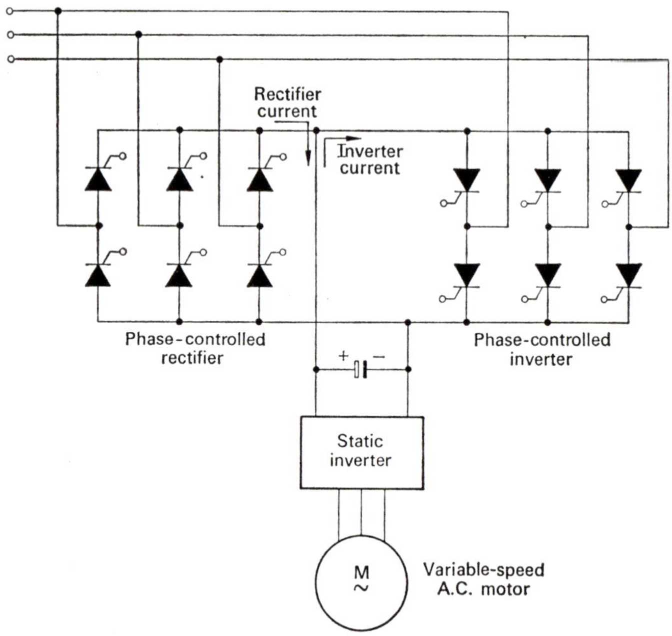

FIG. 3.27. Double-converter connection for regenerative braking of a solid-state a.c. motor drive.

If the d.c. link converter has no provision for regenerative or dynamic braking, the regenerated machine energy causes overcharging of the d.c. link capacitor. This raises the d.c. link voltage and also the terminal voltage of the a.c. motor, causing magnetic saturation and overfluxing of the motor. As a result, there is a large increase in the machine losses, and this provides a certain degree of dynamic braking without any increase in circuit complexity. However, full regeneration would produce an excessive rise in voltage and this method of braking is not suitable for drives which must provide rapid deceleration.

CHAPTER 4

The Phase-controlled Cycloconverter

4.1. INTRODUCTION

In a cycloconverter, the alternating voltage at supply frequency is converted directly to a lower frequency without any intermediate d.c. stage. The operating principles were developed in the 1930s when the grid-controlled mercury-arc rectifier became available. The techniques were applied in Germany, where the three-phase 50-Hz supply was converted to single-phase a.c. at $16\frac{2}{3}$ Hz for railway traction.(84,85) In the United States, a 400-h.p. scheme in which a synchronous motor was supplied from a cycloconverter comprising eighteen thyratrons was in operation for several years as a power station auxiliary drive.(86,87) However, these early schemes were not sufficiently attractive technically or economically and were discontinued.

In recent years the invention of the thyristor and the development of reliable transistorized control circuitry has led to a revival of interest in the cycloconverter principle. Sophisticated control circuits permit the conversion of a fixed input frequency to a variable output frequency at variable voltage, and such schemes are attractive for a.c. motor drives.(39–45) The ruggedness and low weight of the solid-state cycloconverter also make it attractive for aircraft electrical systems, which require the production of a constant output frequency from a variable-speed alternator.(88,89)

4.2. BASIC PRINCIPLES OF OPERATION

The cycloconverter consists of a number of phase-controlled rectifier circuits connected to an a.c. supply system which provides the voltages necessary for delayed phase commutation. The individual circuits are controlled so that a low-frequency output voltage waveform is fabricated from segments of the polyphase input voltages.

Consider a three-phase, half-wave, phase-controlled rectifier supplying an inductive load which maintains continuous current flow and, for simplicity, neglect thyristor forward voltage drop and commutation overlap. With these simplifications, it is shown in Chapter 2 that the average direct voltage output is given by equation (2.2):

$$V_d = V_{do} \cos \alpha$$

where α is the firing angle or delay angle, and V_{do} is the mean output voltage with zero firing delay. This ability to control the output voltage by delayed commutation is the basic property on which phase-controlled cycloconverter operation is based.

Assume the rectifier firing angle is slowly varied as shown in Fig. 4.1. At point A there is zero delay, and the mean output voltage has its maximum value, V_{do}. At B the output voltage is slightly reduced by the introduction of a small firing delay. Further reduction is obtained at C, D and E, while at F the firing angle is $\pi/2$ and V_d is zero. Thus, if the gating

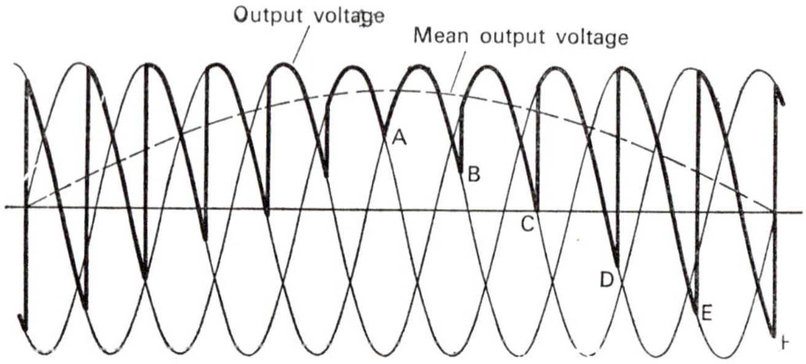

FIG. 4.1. Sinusoidal variation of the mean output voltage of a phase-controlled rectifier.

circuitry is suitably designed, a low-frequency sinusoidal variation may be superimposed on the output voltage, V_d. In Fig. 4.1 the rectifier is shown conducting during intervals of negative output voltage. This means that the circuit is temporarily inverting, and returning reactive energy from the load to the a.c. supply. However, for firing angles between zero and $\pi/2$ the net power flow is from the a.c. supply into the load.

The average back e.m.f. of an inverter may also be controlled in a sinusoidal manner by suitable variation of the firing angle between $\pi/2$ and π. In Fig. 4.2, the back e.m.f. has its maximum value, $-V_{do}$, at point K where $\alpha = \pi$. By reducing α, the back e.m.f. is also reduced as shown at points L, M, N, O, P and Q. For inverter operation, the net power flow is into the a.c. supply, and a voltage must be available to force current flow against the inverter back e.m.f. In the cycloconverter, this voltage is provided by the induced e.m.f. in a reactive load, or by regenerative operation of an a.c. motor.

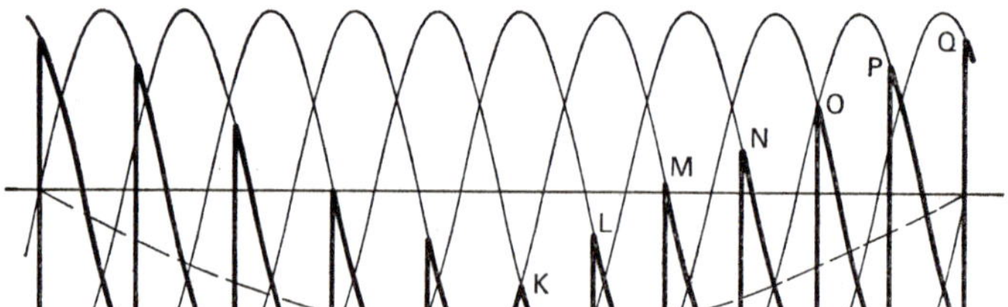

FIG. 4.2. Sinusoidal variation of the mean back e.m.f. of a phase-controlled inverter.

If the firing angle is varied from zero to π and back again to zero, one complete cycle of the low-frequency variation is superimposed on the average output voltage. The superimposed frequency is determined solely by the rate of variation of α and is independent of the supply frequency. Figure 4.3 shows the production of a complete cycle of the low-frequency waveform. This diagram emphasizes the fact that the cycloconverter is basically a switching arrangement. Each thyristor switch opens and closes at suitable instants so that a low-frequency output waveshape is fabricated from segments of the input waveform. The harmonic content of the output voltage decreases as the ratio of output to input frequency is reduced and as the number of supply phases is increased.

The average output voltage of a phase-controlled rectifier can, therefore, be varied sinusoidally through a complete cycle by suitable variation of the delay angle. However, current can only flow in one direction through the circuit and, in order to produce a complete cycle of low-frequency current, two similar circuits must be connected in inverse

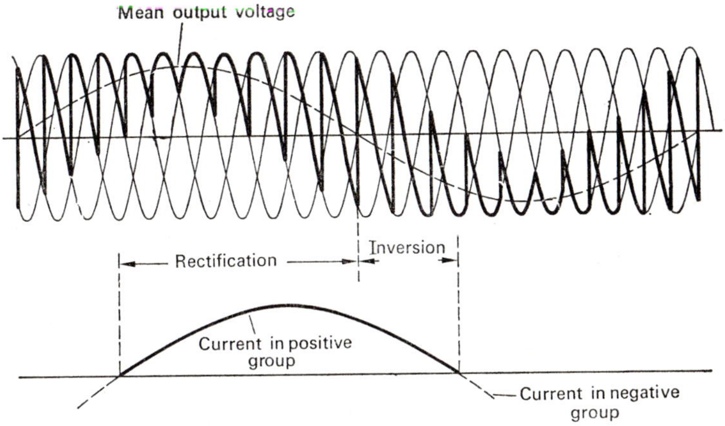

FIG. 4.3. Voltage and current waveforms for the positive group of a phase-controlled cycloconverter when feeding an inductive load at 0·6 power factor.

parallel. The positive rectifier group permits current flow during the positive half-cycle of the low-frequency output wave, while the negative group allows current flow during the negative half-cycle. The arrangement is shown schematically in Fig. 4.4(a), where the three-phase supply produces a single-phase low-frequency output. Any rectifier configuration may be used in the positive and negative group circuits. The simplest arrangement uses two three-phase half-wave circuits as in Fig. 4.4(b).

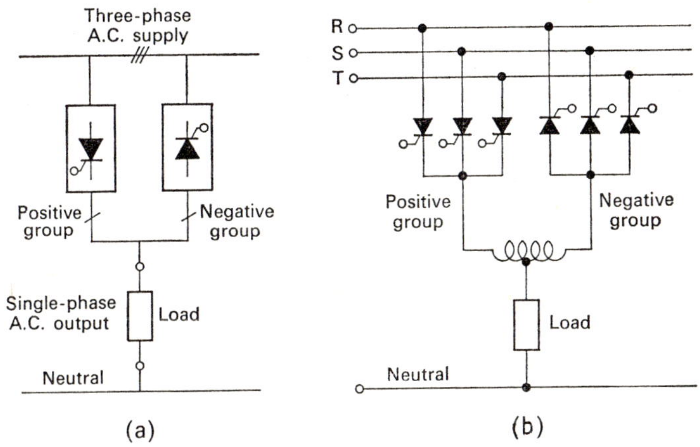

FIG. 4.4. Three-phase to single-phase cycloconverter: (a) schematic diagram; (b) basic circuit.

Since the positive and negative groups are connected in inverse parallel, their average output voltages must always be equal in magnitude and opposite in sign, in order to avoid large circulating currents at output frequency. This is achieved when the delay angles of the positive and negative groups, α_P and α_N, are related by the formula $\alpha_P = \pi - \alpha_N$. However, the instantaneous output voltages of the two groups are quite different, and large harmonic currents will circulate around the low-impedance circuit unless they are limited by using a centre-tapped reactor to combine the outputs, as shown in Fig. 4.4(b). Alternatively, the circulating currents may be suppressed by removing the gating signals from each group for the half-cycle during which the group is not delivering load current. The elimination of circulating currents is examined in more detail in Section 4.3.3.

When a three-phase output is required, three single-phase cycloconverters with a phase displacement of 120° between their outputs are connected as shown in Fig. 4.5(a). With a balanced load, the neutral connection is no longer necessary and may be omitted. The simplest arrangement using three-phase half-wave circuits is shown in Fig. 4.5(b). This circuit requires eighteen thyristors, but if each group consists of a three-phase bridge or six-phase circuit, then thirty-six thyristors are required.

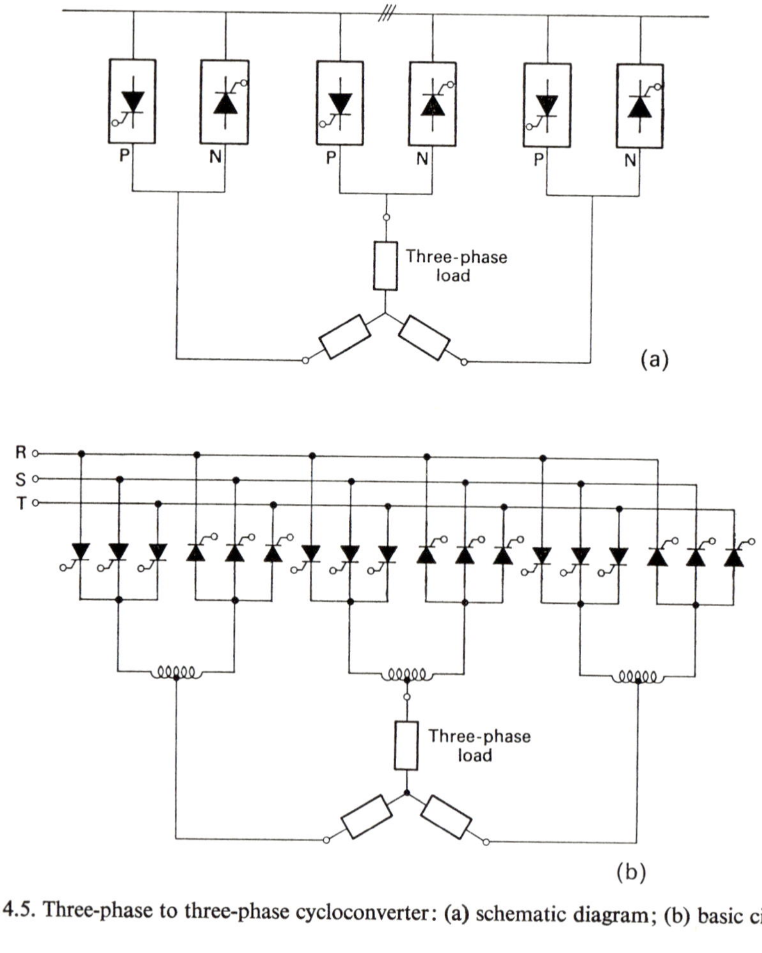

FIG. 4.5. Three-phase to three-phase cycloconverter: (a) schematic diagram; (b) basic circuit.

It has been assumed above, that the output frequency of the cycloconverter is less than the supply frequency. However, the cycloconverter is also capable of operation as a step-up frequency changer, but the output power is limited and circuit losses are high. In practice, with the basic cycloconverter circuit of Fig. 4.5(b), reasonable power output and efficiency are only obtained in the step-down region with output frequencies from zero to about one-third of the input frequency. As the cycloconverter output frequency approaches the supply frequency, the harmonic distortion in the output voltage increases, since the output voltage waveform is composed of fewer segments of the supply voltage. As a result, the losses in the cycloconverter and in the a.c. motor become excessive, and there is a drop in overall efficiency. By using more complex rectifier circuits, the output voltage waveform is improved, and the maximum useful ratio of output to input frequency is increased to about one-half. The a.c. motor normally presents a high impedance at the ripple frequency, and hence the output current is nearly sinusoidal and no additional filtering is necessary.

4.2.1. *Operation with Reactive Loads*

In a cycloconverter, the positive and negative rectifier groups must each supply one-half-cycle of the low-frequency output current. When feeding a resistive load there is no necessity for inverter operation, since the positive group supplies load current during the positive half-cycle of output voltage, and the negative group conducts for the negative half-cycle. When inductive loads are being supplied, the low-frequency output current lags the output voltage. This means that each rectifier group must continue to conduct after its output voltage changes sign. During this period, the group functions as an inverter, and power is returned to the a.c. supply. Inverter operation continues until such time as the load current reduces to zero and reverses, when the other group starts to conduct. In this manner, energy can be transferred in either direction through the cycloconverter.

An interchange of energy between load and source occurs in any simple a.c. circuit with a reactive load (Fig. 4.6). The instantaneous output power is the product of the instantaneous values of load current and voltage. During part of the a.c. cycle, the instantaneous

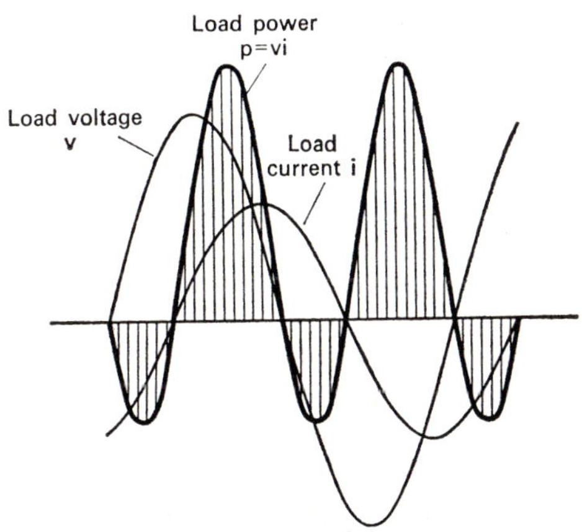

FIG. 4.6. Instantaneous voltage, current and power in a single-phase a.c. circuit.

power is positive and the load absorbs energy from the supply. Later in the cycle, the instantaneous power is negative, and energy is returned to the supply from the magnetic field of the load inductance (or from the electric field of the load capacitance). The difference between the two energies is the net energy delivered to the load. Real power and energy are associated with the component of current which is in phase with the applied voltage, and reactive power is associated with the current component which is 90° out of phase with the applied voltage. Thus, reactive power represents real power which oscillates between source and load and has an average value of zero.

The flow of reactive power associated with an inductive load occurs directly in an a.c. circuit, but when a cycloconverter separates the load from the supply, the reactive power transfer must take place through the cycloconverter. At certain parts of the low-frequency output cycle this demands inverter operation, which occurs automatically whenever the voltage and current of a rectifier group have opposite signs. Thus, the cycloconverter is able to deliver low-frequency a.c. to any type of reactive load, and the phase displacement between the half-cycles of current conduction and output voltage determines the periods of rectifier and inverter operation. In Fig. 4.3 a sinusoidal output current of 0·6 power factor is assumed, and the periods of rectification and inversion are indicated for the positive group. The flow of reactive power is considered further in Section 4.3.2.

4.3. CYCLOCONVERTER CIRCUIT RELATIONS

The behaviour of the phase-controlled cycloconverter circuit will now be examined in more detail.

4.3.1. *Voltage Equation*

The average direct voltage output of a half-wave phase-controlled rectifier with zero firing delay is given by equation (2.1):

$$V_{do} = \sqrt{2}\, V_{ph} \left(\frac{m}{\pi}\right) \sin\left(\frac{\pi}{m}\right)$$

where V_{ph} is the r.m.s. phase voltage and m is the number of secondary phases. As already explained, this result assumes instantaneous commutation and negligible thyristor forward voltage drop.

If the cycloconverter delay angle is slowly varied, the output voltage per phase at any point of the low-frequency cycle may be calculated as the average output voltage for the appropriate delay angle. This ignores the rapid fluctuations superimposed on the average low-frequency waveform. Assuming continuous current conduction, the average output voltage is therefore given by equation (2.2):

$$V_d = V_{do} \cos \alpha .$$

If V_o is the r.m.s. output voltage per phase of the cycloconverter, then the peak output voltage corresponding to zero delay is

$$\sqrt{2}\, V_o = V_{do} = \sqrt{2}\, V_{ph} \left(\frac{m}{\pi}\right) \sin\left(\frac{\pi}{m}\right)$$

and

$$V_o = V_{ph}\left(\frac{m}{\pi}\right)\sin\left(\frac{\pi}{m}\right).$$

However, the firing angle of the positive group cannot be reduced to zero, as this corresponds to a firing angle of π in the negative group ($\alpha_P = \pi - \alpha_N$). In practice, inverter firing cannot be delayed by 180°, since sufficient margin must be allowed for commutation overlap and thyristor recovery time. Consequently, the delay angle of the positive group cannot be reduced below a certain finite value, α_{min}. The maximum output voltage per phase is, therefore,

$$V_{d\,max} = V_{do}\cos\alpha_{min} = rV_{do}$$

where $r = \cos\alpha_{min}$, and is called the voltage reduction factor. The expression for the r.m.s. phase voltage delivered by the cycloconverter is therefore modified to

$$V_o = r\left[V_{ph}\left(\frac{m}{\pi}\right)\sin\left(\frac{\pi}{m}\right)\right]. \tag{4.1}$$

Since α_{min} is necessarily greater than zero, the voltage reduction factor r is always less than unity. By deliberately increasing α_{min} and thereby reducing the range of variation of α about the 90° value, the output voltage V_o can be reduced and a static method of voltage control is obtained. In practice, the output voltage is less than the theoretical value given by equation (4.1), due to commutation overlap and the circulating currents between positive and negative groups.

4.3.2. *Displacement Factor*

The cycloconverter consists essentially of a number of phase-controlled rectifiers operating with a variable firing angle. As already explained in Section 2.5.4, the a.c. supply currents are non-sinusoidal in a rectifier installation, and one must distinguish between the power factor, the displacement factor and the distortion factor. The power factor λ, in such cases, is defined as the ratio of the real input power in watts to the total apparent power in volt-amps. The displacement factor, $\cos\phi$, is the fundamental power factor, since ϕ is the phase displacement between the fundamental phase current and the sinusoidal phase voltage. The distortion factor, μ, is the ratio of the fundamental r.m.s. current to the total r.m.s. current. As shown in Section 2.5.4, these quantities are related by equation (2.6): $\lambda = \mu\cos\phi$. If the commutation overlap between phases is negligible, the displacement factor, $\cos\phi$ equals $\cos\alpha$, where α is the firing angle. Thus, when α is 90°, the displacement factor is zero. In practice, the presence of overlap tends to increase ϕ and so reduces the displacement factor and power factor of the system.

In a cycloconverter, voltage control is obtained by reducing the variation of the delay angle about the 90° value. At low output voltages, the average phase displacement between input current and voltage is large, and the cycloconverter has a low displacement factor. The input current always lags the supply voltage, since phase delay is always present, irrespective of the nature of the load. The cycloconverter cannot transmit leading reactive power, and the lagging reactive power drawn from the supply is always greater than that delivered to the load. A capacitive load consumes leading reactive power but this appears as a demand for lagging reactive power on the input side of the cycloconverter. Thus, the displacement factor has its greatest value when the load is purely resistive, and a capacitive load with a given leading power factor will reduce the displacement factor by exactly the same amount as an inductive load with the same lagging power factor.

In the classical analysis by von Issendorff,[90] the cycloconverter is assumed to have an infinite number of input phases and negligible commutating reactance. When the output voltage has its maximum value, the following expressions are obtained for the reactive power consumption and the input displacement factor:

$$Q_i = Q_o \left[\frac{2}{\pi} \cdot \frac{\cos \phi_o + \phi_o \sin \phi_o}{\sin \phi_o} \right] \tag{4.2}$$

and

$$\cos \phi_i = \frac{1}{\sqrt{\left[1 + 4 \left\{ \dfrac{\cos \phi_o + \phi_o \sin \phi_o}{\pi \cos \phi_o} \right\}^2 \right]}} . \tag{4.3}$$

In these equations, $\cos \phi_i$ is the input displacement factor of the cycloconverter and $\cos \phi_o$ is the output or load power factor; Q_i is the reactive power supplied to the cycloconverter and Q_o is the reactive power consumption of the load. It has been shown that equations (4.2) and (4.3) are also valid for any finite number of input phases.[89]

When voltage control is incorporated in the cycloconverter, the displacement factor is reduced, and in Fig. 4.7 the displacement factor is plotted against the load phase angle for different values of r, the voltage reduction factor. The displacement factor has a maxi-

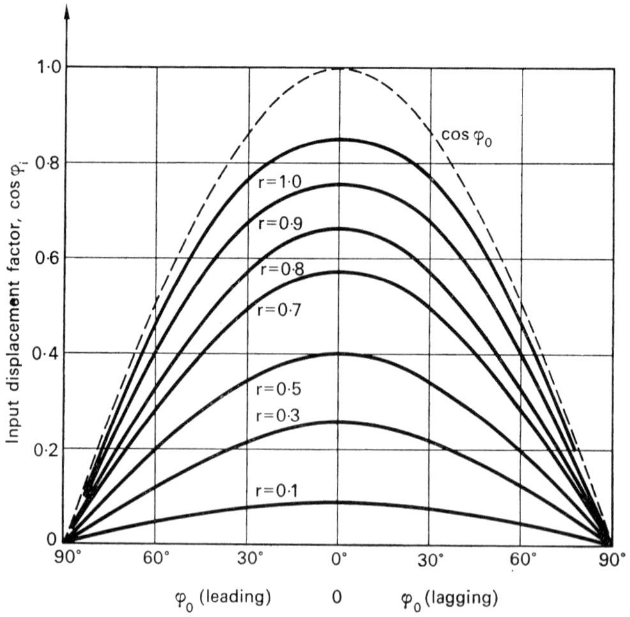

FIG. 4.7. Variation of input displacement factor for a phase-controlled cycloconverter.

mum value of 0·843 with a unity power factor load and maximum output voltage. In practice, as already explained, the voltage reduction factor r is always less than unity, and the maximum displacement factor is somewhat less than the theoretical value. Figure 4.7 also confirms that lagging and leading loads of the same power factor cause similar reductions in the input displacement factor.

The classical analysis can also be used to prove that the input power factor is given by

$$\lambda = \frac{r}{\sqrt{2}} \cos \phi_o \tag{4.4}$$

and the distortion factor is, therefore,

$$\mu = \frac{\lambda}{\cos \phi_i} = \frac{r}{\sqrt{2}} \frac{\cos \phi_o}{\cos \phi_i}. \tag{4.5}$$

4.3.3. *Circulating Currents*

As explained in Section 4.2, the cycloconverter delivers a low-frequency alternating current to each phase of the load through two anti-parallel thyristor groups. The positive group has a delay angle α_P and delivers positive current to the load. The negative group has a delay angle α_N and permits current flow in the opposite or negative direction. If the forward voltage drop of a thyristor is negligible, the delay angles must be controlled so that $\alpha_P = \pi - \alpha_N$. In this manner, the mean output voltage of the rectifier group is maintained equal to the mean back e.m.f. of the inverter group, and the circulation of large low-frequency currents between groups is avoided. However, the instantaneous voltages of the two groups are not identical, and large harmonic currents will circulate unless they are limited or suppressed. These intergroup currents are undesirable, since they increase the circuit losses and impose a heavier loading on the thyristors. They also reduce the displacement factor of the system. The flow of harmonic current may be reduced by the introduction of an intergroup reactor, or completely suppressed by removing the gating pulses from the non-conducting group. These two techniques will now be examined.

Limitation by intergroup reactor. The reactor is connected between the two groups in order to limit the flow of harmonic current, and the load is connected to a centre-tap as in Fig. 4.4(b). The low-frequency output current of the cycloconverter is opposed by the reactance X due to one-half of the intergroup reactor. If the two halves of the reactor are tightly coupled, the flow of harmonic current between groups is opposed by a reactance of $4kX$ where k is the order of the harmonic. A suitable choice of inductance will restrict the flow of harmonic current without seriously affecting the fundamental output current.

In Fig. 4.8 the voltage and current waveforms are shown for the three-phase half-wave anti-parallel connection of Fig. 4.4(b), assuming a highly inductive load. The positive group operates as a rectifier with a delay angle, α_P, which is less than $\pi/2$, while the negative group operates in the inverter region with a delay angle $\alpha_N = \pi - \alpha_P$. Thus, the mean voltages of the two groups are equal, but the instantaneous values are quite different, and the voltage difference appears across the intergroup reactor and has the waveform shown in Fig. 4.8(c).

The circulating current flows through the two series-connected thyristor circuits and is limited only by the inductance of the intergroup reactor, assuming negligible circuit resistance. The circulating current is therefore proportional to the integral of the intergroup voltage and has the variation shown in Fig. 4.8(d). In the three-phase half-wave system under consideration, the maximum circulating current occurs when the firing angle α_P is 60°, and α_N is 120°. During the intervals when the positive and negative thyristor groups have different instantaneous voltages, the intergroup reactor behaves as a potential divider, and the ouput voltage at the centre-tap is the average of the two group voltages, as shown in Fig. 4.8(e).

In a cycloconverter application, the delay angles α_P and α_N are continuously modulated, and the output voltage waveform passes smoothly through each of the stages shown in Fig. 4.8(e). The output waveform is considerably improved if a six-phase system is used. The intergroup voltages and circulating currents are then significantly reduced, and maximum circulating current occurs with a delay angle of 90°, corresponding to zero output voltage.[91,92]

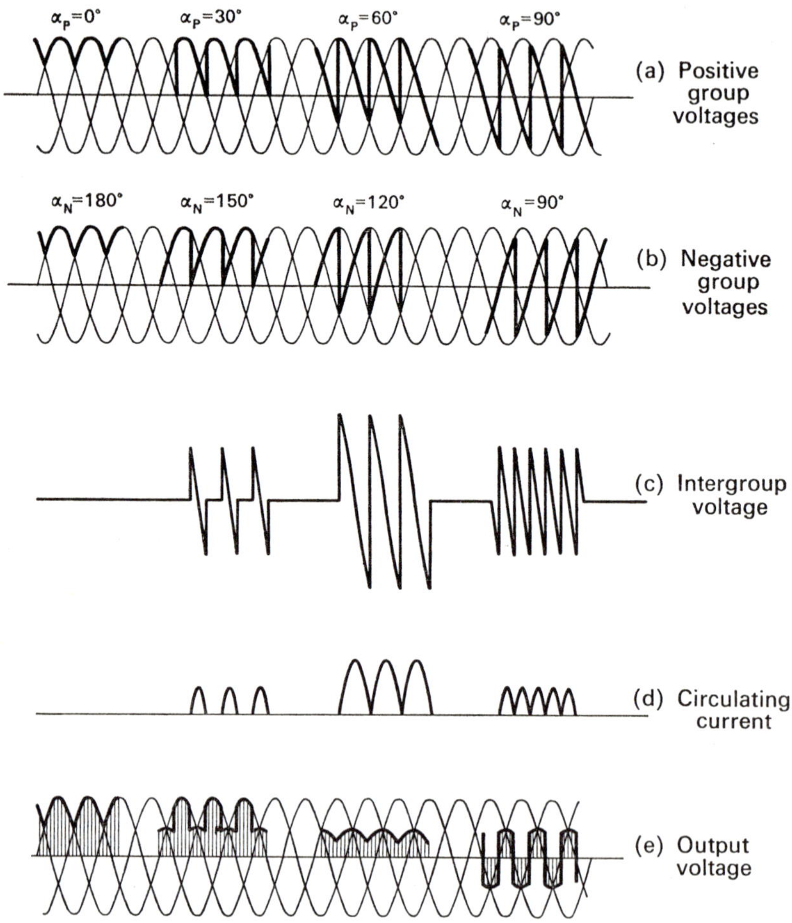

FIG. 4.8. Voltage and current waveforms for the three-phase half-wave anti-parallel connection of Fig. 4.4(b).

Suppression by intergroup blanking. In modern versions of the cycloconverter, the circulating current is usually suppressed by blocking all thyristors in the rectifier group which is not delivereing load current. This is achieved by removing the gating pulses for the appropriate periods, and the intergroup reactors of Fig. 4.5(b) can then be reduced in size or completely eliminated. A current-sensing device is incorporated in each output phase of the cycloconverter. This detects the direction of the output current and feeds a signal to the control circuitry to inhibit or "blank" the gating of thyristors in the non-conducting group. If an overload or fault current flows in the system, all gating pulses are removed in order to protect the thyristors.

The suppression of the circulating current by intergroup blanking improves the efficiency and displacement factor of the cycloconverter, and also increases the maximum usable output frequency. On resistive loads, the cycloconverter output voltage has the discontinuous waveform shown in Fig. 4.9, and the effect of the intergroup blanking is evident. The positive group delivers voltage and current to the load for the positive half-cycle, and the negative group determines the voltage and current during the negative half-cycle. Since only one rectifier group is gated at a time, the load voltage is the same as the output voltage of the conducting group, and the potential-divider effect of Fig. 4.8 is not present. If the load is sufficiently inductive, the output current is continuous and lags the voltage. As the load current passes through zero, it transfers from one rectifier group to the other. Since $\alpha_P + \alpha_N = \pi$, the incoming group delivers the same voltage as the outgoing group, and the load voltage also transfers smoothly from one group to the other.

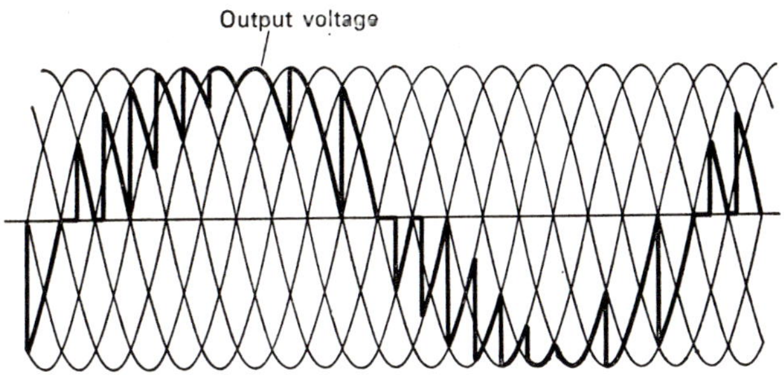

FIG. 4.9. Output voltage waveform for a phase-controlled cycloconverter with intergroup blanking and a resistive load.

4.3.4. *Cycloconverter Analysis*

The asymmetrical nature of the thyristor results in non-sinusoidal current and voltage waveforms in phase-controlled cycloconverter circuits. In order to simplify the analysis, the classical derivation of the circuit relations by von Issendorff[90] makes a number of simplifying approximations. An infinite number of input phases is assumed, and a highly inductive load maintains continuous current conduction. The circulating currents between rectifier groups are limited by intergroup reactors, and intergroup blanking is not employed. In present-day practical cycloconverters,[93] these circuit conditions and approximations are often invalid and an exact analysis becomes extremely complicated. The difficulty may be avoided by making other appropriate circuit approximations in order to obtain equivalent circuits for the purposes of analysis.[94,95] Alternatively, the thyristor may be simulated by an analogue computer circuit using operational amplifiers and reed relays.[96] Several of these units are then interconnected to simulate cycloconverter operation. The analogue computer simulation allows convenient and rapid variation of the circuit parameters so that practical cycloconverter systems may be designed and optimized.

4.4. FREQUENCY AND VOLTAGE CONTROL

An ideal phase-controlled rectifier in which current flow is continuous, delivers an output voltage waveform with an average value of zero when the firing delay, α, is $\pi/2$. If the delay angle is biased to the $\pi/2$ position, and is slowly varied by $\pm\pi/2$ on either side of this point, a low-frequency output voltage is obtained with peak values of $\pm V_{do}$. When the range of oscillation of α on either side of the 90° value is reduced, the magnitude of the output voltage is also reduced, and voltage control is obtained. The cycloconverter output frequency is determined by the rate at which α is varied about the 90° point, and hence output voltage and frequency are both governed by the gating pulses from the control circuitry.

The repetition frequency and sequence of the gating pulses are determined by the cycloconverter a.c. supply voltages and a reference or control voltage which is generated in a low-power master oscillator. This oscillator has the same number of phases as the cycloconverter output, and generates reference voltages with the waveshape, frequency and phase sequence required in the output. The supply voltages and reference voltages are fed to a comparator circuit, and the output signals control the gating circuits. The gating signals are designed to duplicate the reference conditions in the cycloconverter output voltages. Normally the reference voltages are sinusoidal, and the gating pulses modulate the various delay angles so that the average output voltage has a superimposed sinusoidal variation. A positive reference voltage advances the firing of the positive group and delays the firing of the negative group by the same amount in order to satisfy the relationship $\alpha_P + \alpha_N = \pi$. Variation of the reference oscillator frequency produces a corresponding variation in the output frequency in an open-loop system similar to that already described for the d.c. link converter. The advantages of stability, accuracy and fast transient response are again present. The direction of rotation of the drive motor is reversed by simply changing the phase sequence of the reference oscillator, and no switching is necessary in the power circuits. The amplitude of the cycloconverter output voltage is proportional to the reference voltage amplitude and hence the cycloconverter may be regarded as a linear high-gain power amplifier. Independent control of voltage and frequency is possible, and maximum output voltage may be obtained at any operating frequency. Alternatively, the required volts/Hz variation may be programmed into the control circuitry.

4.4.1. *Control Circuitry*

The simplest method of generating the reference frequency signal is by means of an electromechanical system. A synchro, with a single-phase rotor and three stator windings 120° apart, is supplied with rotor excitation at supply frequency, and voltages of this frequency are induced in each stator winding. The rotor is driven by a small reference motor at a speed proportional to the required output frequency, and hence the stator voltages are modulated at output frequency. The three-phase, low-frequency reference voltages are obtained by demodulating the stator voltages. These are then fed to the transistorized logic and gating circuits. The cycloconverter output frequency is controlled by variation of the synchro speed, and a reversal of the main motor is obtained by reversing the reference motor. The magnitude of the cycloconverter output voltage is varied independently by controlling the synchro excitation.

Completely static reference oscillators have replaced these electromechanical systems, giving increased reliability and a reduction in size. A three-stage resistance–capacitance phase-shift oscillator will generate the required three-phase reference voltages but, in order to obtain a variable-frequency output down to d.c., the beat-frequency output from two similar high-frequency oscillators must be used. One of these oscillators has an adjustable output frequency which can be controlled by a d.c. reference voltage, whose polarity determines the output phase sequence.

As already pointed out, a delay angle of $\pi/2$ corresponds to zero output voltage from a cycloconverter group, and modulation of α about this value produces an average output voltage $V_d = V_{do} \cos \alpha$. If an origin is taken at $\alpha = \pi/2$ by defining a new variable, $\psi = \pi/2 - \alpha$, then a variation of α from 0 to π corresponds to a variation of ψ from $+\pi/2$ to $-\pi/2$, and the mean output voltage is $V_d = V_{do} \sin \psi$. Thus, if ψ is varied uniformly between the limits $+\pi/2$ and $-\pi/2$, the output voltage has a sinusoidal variation with an amplitude of V_{do}. The variation in ψ as a function of time has, therefore, a symmetrical triangular waveform, and by reducing its amplitude, the output voltage is reduced, but the sinusoidal output waveshape is preserved.

The low-frequency sinusoidal output of the reference oscillator must be translated into a sinusoidal variation of the cycloconverter output voltage by means of the triangular waveshape for the angle ψ. This is achieved in the mercury-arc cycloconverter by using the bias-shift method of grid-control. In a polyphase mercury-arc rectifier or inverter using bias-shift control, sinusoidal supply-frequency grid voltages of constant amplitude are obtained from the secondary windings of a small grid-excitation transformer. A variable d.c. bias voltage is introduced between the star-point of the grid-transformer secondary and the common cathode terminal, and is thereby superimposed on the sinusoidal firing voltages. Each secondary phase of the grid transformer supplies one grid of the mercury-arc rectifier, and valve conduction is initiated when the grid-cathode voltage goes positive and exceeds a certain critical value. Phase control is obtained by variation of the d.c. bias potential, since this determines the instant at which the net voltage on the grid exceeds the critical firing value.

In a mercury-arc cycloconverter group, the low-frequency sinusoidal reference voltage replaces the d.c. bias voltage, and thus a continuous variation in the firing angle is obtained. In a three-phase half-wave circuit, the supply-frequency grid voltage component should lag the corresponding anode voltage by 120°. If this condition is satisfied, a sinusoidal variation in the low-frequency reference voltage produces a linear variation in ψ about the $\pi/2$ value and, as a result, the mean cycloconverter output voltage has the desired sinusoidal waveshape.[62] This technique is readily adapted to modern thyristor circuits by amplifying and squaring the resulting grid voltage waveforms and using the positive- and negative-going edges to generate suitable thyristor gating pulses. Information regarding the output current of the cycloconverter may also be fed to the control circuitry in order to inhibit firing of the non-conducting rectifier group.

In the foregoing analysis it has been assumed that a sinusoidal output voltage is required, but the control circuitry is readily modified so that other output waveshapes are obtained.[97] A square-wave output voltage has a minimum of phase delay and hence reduces the reactive power consumption to a minimum. It is therefore advantageous to use square-wave voltages in applications where such waveforms are acceptable. However, when a sine-wave output is required, it is not feasible to generate a square waveform and then filter the output to obtain the fundamental component, since this usually results in an over-sized and uneconomic filter.

4.5. COMPARISON OF THE CYCLOCONVERTER AND D.C. LINK CONVERTER

The cycloconverter and d.c. link converter both deliver an alternating voltage at a frequency which is determined by the reference oscillator. This technique provides an extremely stable and accurate output frequency, which is independent of fluctuations in the supply frequency and voltage. The frequency control is an open-loop system with no closed-loop stability problems, and speed reversal is readily obtained by reversing the thyristor firing sequence. In the cycloconverter, voltage control is also obtained by means of the control circuitry. A completely static supply is therefore available, with a transient response in the millisecond range. This is also provided by d.c. link systems which employ static voltage control techniques such as phase-shift control or pulse-width modulation.

The cycloconverter has some advantages and some disadvantages compared with the d.c. link converter. The following advantages can be listed:

1. In a cycloconverter, a.c. power at one frequency is converted directly to a.c. at a lower frequency in a single conversion stage. The d.c. link converter has two power controllers in cascade, and the full output power is converted twice.

2. The cycloconverter functions by means of phase commutation, and no auxiliary forced-commutating circuits are necessary. This results in a more compact power circuit and also eliminates the circuit losses associated with forced commutation. An overall drive efficiency of 90 per cent has been achieved using a squirrel-cage induction motor and a cycloconverter comprising thirty-six thyristors with intergroup blanking.[44]

3. The cycloconverter is inherently capable of power transfer in either direction between source and load, and can therefore supply a.c. power to loads of any power factor. It is also capable of regenerative operation at full power over the complete speed range down to standstill. This feature is difficult to incorporate in static inverter drives, and hence the cycloconverter is preferable for large reversing drives requiring rapid acceleration and deceleration. This type of application occurs principally in the metal rolling industry.[42,45]

4. In the cycloconverter, a commutation failure causes a short-circuit of the a.c. supply, but if an individual thyristor fuse blows, a complete shut-down is not necessary, since the cycloconverter continues to function with a somewhat distorted output waveform. A balanced load is presented to the a.c. supply even with unbalanced output conditions.

5. The cycloconverter delivers a high-quality sinusoidal waveform at low output frequencies, since the low-frequency wave is fabricated from a large number of segments of the supply waveform. On the other hand, many static inverters generate a stepped-wave voltage which may cause a jerky rotation of the a.c. motor at frequencies below 10 Hz. The distorted waveform also slightly accentuates the danger of system instability at low frequencies (Chapter 5). For these reasons, the cycloconverter is often preferable for very-low-speed applications.

The cycloconverter also has the following disadvantages:

1. The principal disadvantage is the fact that the maximum output frequency must be less than about one-third or one-half of the input frequency for reasonable power output and efficiency. This is a serious limitation when operating on 50 or 60 Hz a.c. supplies, as it restricts the motor speed to a maximum of approximately 3000 rev/min even with a two-pole machine. This disadvantage can be overcome in applications where the a.c. power is generated by an engine-driven alternator and used directly to drive the a.c. motors. By selecting a high alternator frequency, high maximum speeds can be obtained with compact high-frequency a.c. motors.

2. The cycloconverter requires a large number of thyristors, and its control circuitry is more complex than that employed in many d.c. link converters. This expensive circuitry is not justified in small installations, but the cycloconverter is economic for units of 20 kVA or more. It may be advantageous to use two-phase motors in order to reduce the number of thyristors and improve thyristor utilization.

3. The cycloconverter has a low input power factor, particularly at reduced output voltages. In a d.c. link converter, a high input power factor may be obtained for all operating conditions by using a simple diode rectifier input.

To summarize, this comparison of the cycloconverter and d.c. link converter indicates that the d.c. link system is more suitable for high frequencies, but the cycloconverter is extremely attractive for low-speed reversible drives.

CHAPTER 5

A.C. Motor Operation with Non-sinusoidal Supply Waveforms

5.1. INTRODUCTION

Most static frequency converters generate an output voltage waveform with a significant harmonic content. In this chapter, the motor performance with non-sinusoidal supply voltages is compared with normal sine-wave operation. The fundamental component of the non-sinusoidal voltage is assumed to have the same amplitude as the rated sine-wave voltage of the machine.

5.2. AIR-GAP M.M.F. HARMONICS

The elementary two-pole three-phase machine has three stator coils displaced 120° apart in space, and excited by a three-phase system of currents which are 120° apart in time. In normal operation, each coil carries a sinusoidal current which establishes a pulsating magnetic field or magnetomotive force (m.m.f.). It may be assumed that each phase establishes its m.m.f. wave in the air gap independently of the other two phases. Since the phase winding is usually distributed in a number of slots in an iron surface, the spatial distribution of m.m.f. is non-sinusoidal. However, the actual m.m.f. distribution may be resolved into a fundamental component and a series of higher odd space harmonics. For the present, only the fundamental component is considered, and consequently the air gap of the machine has three sinusoidally distributed m.m.f. waves which are 120° apart in space. As each phase current varies, the corresponding m.m.f. wave pulsates in magnitude but retains its sinusoidal space distribution.

Let F_1 denote the amplitude of the m.m.f. wave due to coil 1 at some instant. The spatial m.m.f. distribution is therefore given by

$$f_1 = F_1 \cos \delta \tag{5.1}$$

where δ represents the angular displacement on the armature surface, with the origin chosen on the axis of coil 1. As the coil current varies sinusoidally, the amplitude F_1 varies proportionately, and a standing m.m.f. wave is produced. The instantaneous amplitude F_1 is given by

$$F_1 = \hat{F}_1 \sin \omega t \tag{5.2}$$

where $\hat{F}_1$ is the maximum value of F_1 corresponding to the peak coil current, and a time origin is taken at the instant when the current in coil 1 is zero.

Combining equations (5.1) and (5.2) gives the spatial distribution of m.m.f. due to coil 1 at time t as

$$f_1 = \hat{F}_1 \cos\delta \sin\omega t. \tag{5.3}$$

The m.m.f. wave due to coil 2 is displaced by 120° in space and time relative to f_1, and hence is given by

$$f_2 = \hat{F}_1 \cos(\delta - 2\pi/3) \sin(\omega t - 2\pi/3).$$

Similarly for coil 3

$$f_3 = \hat{F}_1 \cos(\delta - 4\pi/3) \sin(\omega t - 4\pi/3).$$

The resultant air-gap m.m.f. due to the entire winding is obtained by adding the contributions of the three phases. Thus

$$\begin{aligned} f &= f_1 + f_2 + f_3 \\ &= \hat{F}_1 \{\cos\delta \sin\omega t + \cos(\delta - 2\pi/3)\sin(\omega t - 2\pi/3) + \cos(\delta - 4\pi/3)\sin(\omega t - 4\pi/3)\}. \end{aligned}$$

Using the trigonometric relation that $\cos A \sin B = \frac{1}{2}\sin(A+B) - \frac{1}{2}\sin(A-B)$, the above expression reduces to

$$f = \tfrac{3}{2}\hat{F}_1 \sin(\omega t - \delta) \tag{5.4}$$

This represents a sinusoidal distribution of m.m.f. of constant amplitude, which moves with a uniform angular velocity, ω, in the direction of increasing δ. This may be understood by noting that the sine wave moves through a distance ωt in a time t, and hence the displacement increases uniformly with time. Alternatively, a point P may be considered which moves with the wave so that the m.m.f. at P is constant at all times. For such a point, $\sin(\omega t - \delta)$ is constant, and hence $(\omega t - \delta) =$ constant. On differentiating with respect to t, $\omega - \mathrm{d}\delta/\mathrm{d}t = 0$ or $\mathrm{d}\delta/\mathrm{d}t = \omega$. This verifies the familiar result that a three-phase winding excited by balanced sinusoidal currents produces a fundamental m.m.f. wave rotating at synchronous speed.

If the phase currents have a non-sinusoidal waveform, additional harmonic m.m.f. waves are present in the air gap. The existence of a particular harmonic may be confirmed by an analysis similar to that presented above for the fundamental m.m.f.

5.2.1. *Time Harmonic M.M.F. Waves*

Time harmonic waves of m.m.f. are produced by current harmonics in the phase windings. Assume, for example, that the phase current has a fifth harmonic component, and consequently each phase establishes a standing m.m.f. wave having the same spatial distribution as the fundamental, but pulsating at five times the supply frequency. The fifth harmonic m.m.f. due to coil 1 is therefore given by

$$f_1 = \hat{F}_{1,5} \cos\delta \sin 5\omega t$$

where $\hat{F}_{1,5}$ is the peak amplitude of the fundamental space m.m.f. due to the fifth harmonic of current. Similarly,

$$f_2 = \hat{F}_{1,5} \cos(\delta - 2\pi/3) \sin 5(\omega t - 2\pi/3)$$

and

$$f_3 = \hat{F}_{1,5} \cos(\delta - 4\pi/3) \sin 5(\omega t - 4\pi/3).$$

The resultant m.m.f. is obtained, as before, by adding the three m.m.f. contributions. Thus,

$$f = \tfrac{3}{2}\hat{F}_{1,5} \sin(5\omega t + \delta). \tag{5.5}$$

This confirms that a rotating m.m.f. is produced by the fifth harmonic currents. The speed of rotation is given by $d\delta/dt = -5\omega$. This means that the wave is moving at five times synchronous speed in the opposite direction to the fundamental m.m.f.

Similarly, it can be shown that the seventh harmonic currents produce a rotating m.m.f. wave which moves at seven times synchronous speed in the same direction as the fundamental. In general, current harmonics of order $k = 3n + 1$, where $n = 0, 1, 2, 3, \ldots$, produce forward-rotating m.m.f. waves, while harmonics of order $k = 3n + 2$ produce backward-rotating waves. The speed of rotation of a time harmonic field is always k times the synchronous speed.

5.2.2. *Space Harmonic M.M.F. Waves*

In the analysis above, it has been assumed that each phase current establishes a fundamental space m.m.f. wave, and the presence of higher odd space harmonics has been ignored. In fact, even when the polyphase winding is excited by purely sinusoidal currents, the space harmonic distributions of m.m.f. due to different phases combine to produce rotating harmonic m.m.f. waves. Thus, the fifth space harmonic m.m.f. due to fundamental current in coil 1 may be written:

$$f_1 = \hat{F}_{5,1} \cos 5\delta \sin \omega t.$$

Corresponding expressions are written for f_2 and f_3 and, when the three m.m.f.s are combined, the result is:

$$f = \tfrac{3}{2}\hat{F}_{5,1} \sin(\omega t + 5\delta). \tag{5.6}$$

This result confirms the existence of a fifth space harmonic m.m.f. rotating backwards at one-fifth of synchronous speed. Similarly, the seventh space harmonic can be shown to rotate forwards at one-seventh of synchronous speed. These are the well-known space harmonic waves of m.m.f. produced by a symmetrical integral-slot winding excited by balanced sinusoidal currents.

When harmonic currents are present in the phases of the integral-slot winding, the time and space harmonic m.m.f. waves predicted above are present simultaneously. Additional rotating m.m.f. waves are produced by the space harmonic distributions of m.m.f. due to time harmonics in the phase currents. The existence of a particular harmonic is again confirmed by summing the contributions of the three phases. For example, the fifth harmonic currents in each of the three phases produce the following seventh harmonic space m.m.f.s:

$$f_1 = \hat{F}_{7,5} \cos 7\delta \sin 5\omega t,$$

$$f_2 = \hat{F}_{7,5} \cos 7(\delta - 2\pi/3) \sin 5(\omega t - 2\pi/3),$$

$$f_3 = \hat{F}_{7,5} \cos 7(\delta - 4\pi/3) \sin 5(\omega t - 4\pi/3).$$

On combining these three components

$$f = \tfrac{3}{2}\hat{F}_{7,5} \sin(5\omega t + 7\delta). \tag{5.7}$$

The existence of a seventh space harmonic m.m.f. rotating backwards at five-sevenths of synchronous speed is, therefore, confirmed.

The complete results are summarized in Table 1, where all speeds are expressed as multiples of the synchronous speed which is denoted by +1. A positive sign denotes a wave rotating in the same direction as the fundamental, and a negative sign indicates a backward-rotating harmonic. The first row of the table shows the time harmonic waves of m.m.f. already

TABLE 1. M.M.F. COMPONENTS OF A THREE-PHASE ARMATURE WINDING

Order of space harmonic h	Order of time harmonic, k						
	1	3	5	7	9	11	13
1	+1	—	−5	+7	—	−11	+13
3	—	±1	—	—	±3	—	—
5	$-\frac{1}{5}$	—	+1	$-\frac{7}{5}$	—	$+\frac{11}{5}$	$-\frac{13}{5}$
7	$+\frac{1}{7}$	—	$-\frac{5}{7}$	+1	—	$-\frac{11}{7}$	$+\frac{13}{7}$
9	—	$\pm\frac{1}{3}$	—	—	±1	—	—
11	$-\frac{1}{11}$	—	$+\frac{5}{11}$	$-\frac{7}{11}$	—	+1	$-\frac{13}{11}$
13	$+\frac{1}{13}$	—	$-\frac{5}{13}$	$+\frac{7}{13}$	—	$-\frac{11}{13}$	+1
15	—	$\pm\frac{1}{5}$	—	—	$\pm\frac{3}{5}$	—	—

analysed. These are the fundamental space m.m.f. waves due to harmonics in the phase currents. The first column shows the spatial harmonic waves of m.m.f. due to the fundamental phase currents. The remainder of the table shows the space harmonics up to the fifteenth due to time harmonics up to the thirteenth. Most inverter circuits do not generate even-numbered time harmonics and these are omitted from the table. A similar m.m.f. analysis can be performed for a two-phase a.c. winding.[98]

5.2.3. *Amplitude of M.M.F. Harmonics*

For a particular applied voltage waveform, the harmonic m.m.f. amplitudes relative to the fundamental are determined by the winding arrangement used. For a three-phase winding, the hth space harmonic rotating m.m.f. due to the kth time harmonic of current is readily shown to be:

$$F_{h,k} = 1{\cdot}35 k_{wh} \frac{1}{h} \frac{N_{ph}}{p} I_k \text{ ampere-turns per pole} \tag{5.8}$$

where k_{wh} is the hth space harmonic winding factor,
N_{ph} is the number of series turns per phase,
p is the number of pole-pairs, and
I_k is the r.m.s. value of the kth harmonic current.

For the fundamental rotating m.m.f. wave, $h = k = 1$ and the amplitude is:

$$F_{1,1} = 1{\cdot}35 k_w \frac{N_{ph}}{p} I_1 \quad \text{ampere-turns per pole}$$

where k_w is the fundamental winding factor, and
I_1 is the fundamental r.m.s. phase current.

The harmonic m.m.f. amplitude expressed as a fraction of the fundamental amplitude is:

$$\frac{F_{h,k}}{F_{1,1}} = \frac{k_{wh}}{k_w} \frac{1}{h} \frac{I_k}{I_1}. \tag{5.9}$$

For normal well-designed three-phase motors, the harmonic winding factor k_{wh} is much less than k_w, and the space harmonic m.m.f. waves have negligible amplitudes. This is also true for two-phase motors which are occasionally employed in static a.c. drives in order to limit the amount of inverter circuitry required. In the remainder of this chapter attention is therefore confined to the time harmonic m.m.f. waves which have a fundamental space distribution.

5.2.4. *Positive, Negative and Zero-sequence Harmonics*

As shown in the first column of Table 1, the time harmonic waves of m.m.f. due to harmonic currents of order $k = 3n + 1$ (where n is zero or an integer) rotate in the same direction as the main field, while harmonic currents of order $(3n + 2)$ generate backward-rotating m.m.f. waves. In general, the direction of rotation of the harmonic field is determined by the phase sequence of the harmonic currents. Harmonics of order $(3n + 1)$ have the same phase sequence as the fundamental currents, and hence produce an m.m.f. wave which rotates in the same direction as the main field. Harmonics of order $(3n + 1)$ are, therefore, called positive-sequence harmonics. Conversely, harmonic currents of order $(3n + 2)$ are called negative-sequence harmonics, since they have the opposite phase sequence to the fundamental, and generate backward-rotating m.m.f. waves. Harmonic currents of order $(3n + 3)$ do not produce any fundamental air-gap m.m.f., since they are in exact time phase in each of the three windings (see Table 1). These are termed the zero-sequence currents. Since the harmonic order is always a multiple of three, they are also termed triplen harmonics.

5.3. HARMONIC BEHAVIOUR OF A.C. MOTORS [79]

When an a.c. motor is operated on a non-sinusoidal supply, the stator voltage can be analysed into a fundamental component and a series of harmonics. If magnetic saturation is neglected, the motor may be regarded as a linear device, and the principle of superposition can be applied. This means that the motor behaviour can be analysed independently for the fundamental voltage and for each harmonic term. The overall response to the non-sinusoidal voltage is then obtained as a summation of the responses to the individual components.

Thus, the net motor current or torque is equal to the sum of the current or torque contributions of each voltage component in the supply waveform. It is convenient to express the motor current and torque in per-unit or normalized form, that is, the actual values of current and torque are expressed as fractions of the rated current and torque of the motor.

5.3.1. *Harmonic Equivalent Circuits*

The conventional equivalent circuit for one phase of an induction motor on a sinusoidal supply is shown in Fig. 5.1(a) and is derived in Appendix A. In this circuit, the core losses and saturation effects are neglected, and X_1 and X_2 are the stator and rotor leakage reactances at the supply frequency. X_m is the corresponding magnetizing reactance. The rotor slip with respect to the fundamental rotating field is denoted by s_1, and hence

$$s_1 = \frac{n_1 - n}{n_1} \tag{5.10}$$

where n_1 is the synchronous speed of the rotating field, and n is the actual rotor speed.

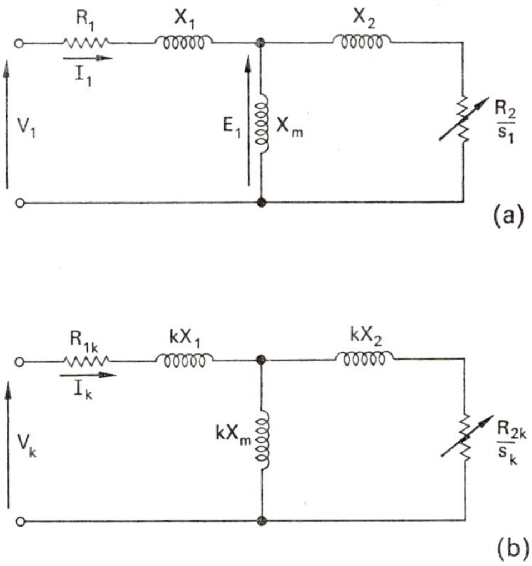

FIG. 5.1. Induction motor equivalent circuit diagrams: (a) fundamental-frequency equivalent circuit; (b) equivalent circuit for the kth time harmonic (positive or negative sequence).

The kth harmonic in the phase currents produces a time harmonic m.m.f. rotating forwards or backwards at a speed kn_1. The rotor slip in a forward-rotating harmonic field is

$$s_k = \frac{kn_1 - n}{kn_1}$$

and for a backward-rotating field

$$s_k = \frac{kn_1 + n}{kn_1}.$$

In general, therefore,

$$s_k = \frac{kn_1 \mp n}{kn_1} \tag{5.11}$$

where the negative sign is valid for positive-sequence harmonics and the positive sign applies to negative-sequence harmonics.

The harmonic slip, s_k, is expressed in terms of s_1 by substituting for n from equation (5.10). This gives

$$s_k = \frac{(k \mp 1) \pm s_1}{k}. \tag{5.12}$$

The fundamental equivalent circuit of Fig. 5.1(a) may be adapted for the kth harmonic voltage and current as shown in Fig. 5.1(b). The harmonic slip, s_k, is substituted for the fundamental slip, s_1, and all inductive reactances are increased by a factor k. The stator and rotor resistances are also larger due to skin effect at the harmonic frequency. Strictly speaking, the rotor leakage inductance is also modified by the skin effect, and this must be taken into consideration in precise calculations.[99]

It may be verified by means of equation (5.12) that there is very little variation in s_k for normal motor operation. If the motor speed varies from synchronous speed to standstill, the fundamental slip, s_1, varies from 0 to 1, but the fifth harmonic slip, s_5, only varies from 1·2 to 1. The corresponding variation in s_7 is 0·857 to 1, and for higher harmonics s_k is even closer to unity. The harmonic equivalent circuit of Fig. 5.1(b) can be simplified as shown in Fig. 5.2(a) by removing the resistances. This is justified by the fact that the inductive

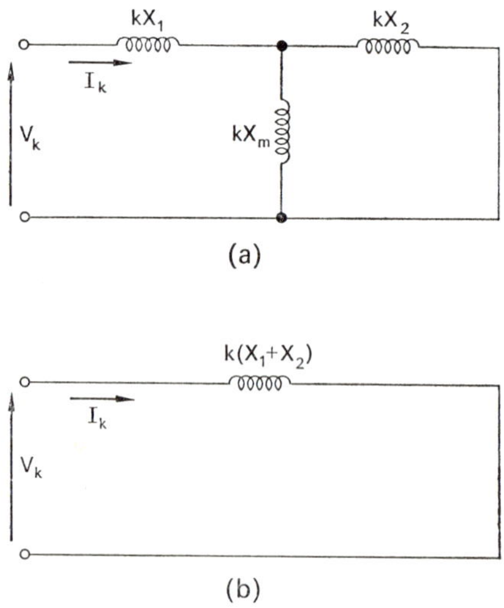

FIG. 5.2. Approximate equivalent circuits for harmonic current calculations.

reactances increase linearly with frequency, while the increase in rotor resistance with frequency due to skin effect is less than linear. Since s_k is approximately unity, the circuit resistance can be neglected in comparison with the reactance at the harmonic frequencies. Further simplification is possible as in Fig. 5.2(b), since the shunt magnetizing reactance is much greater than the rotor leakage reactance, and may be omitted. The motor impedance presented to harmonic currents is, therefore, approximately $k(X_1 + X_2)$ where X_1 and X_2 are the stator and rotor leakage reactances at the fundamental supply frequency.

The zero-sequence stator current harmonics are in time phase, and consequently do not produce a space fundamental rotating m.m.f. wave. However, the zero-sequence currents may establish pulsating space harmonic m.m.f. waves in the air gap, and each pulsating wave can be resolved into a forward and backward-travelling wave, as shown in Table 1. These flux waves induce unequal harmonic currents in the moving rotor, and hence the presence of zero-sequence stator currents can affect the motor torque.

The reactance presented to the flow of the kth zero-sequence harmonic is kX_0, where X_0 is the stator zero-sequence reactance at fundamental frequency. If X_0 is small and the applied voltage has a large zero-sequence component, the resulting zero-sequence currents may cause a significant stator copper loss, and seriously reduce the motor efficiency. However, zero-sequence currents can only flow in a star-connected system having a neutral connection between source and load, as otherwise there is no return path for the in-phase zero-sequence currents. In practice, most inverter circuits do not generate zero-sequence voltages, but if these components are present they are offered an infinite zero-sequence impedance by isolating the neutral connection.

The fundamental equivalent circuit of Fig. 5.1(a) only applies to the polyphase induction motor, but the subsequent harmonic circuits are also valid for the harmonic behaviour of the reluctance or synchronous motor, since these machines operate asynchronously with respect to the time harmonic m.m.f. waves. When analysing motor operation at very low supply frequencies, the approximate harmonic equivalent circuits of Fig. 5.2 may not be valid, since the winding resistance can be a significant factor at the low harmonic frequencies. The simplifications are usually justified, however, if the fundamental frequency exceeds about 10 Hz.

5.3.2. *Harmonic Currents*

Since s_k is nearly unity at all motor speeds from standstill to synchronism, the harmonic equivalent circuit of Fig. 5.1(b) is practically independent of motor speed, and this is emphasized by the approximate circuits derived from it. Thus, the harmonic currents remain constant for all operating conditions of the motor from no-load to full-load, and even down to stand still. The fundamental stator current is determined by the motor loading and, as a result, the relative harmonic content of the machine current is considerably greater for light-load operation than for full-load or starting conditions. This causes a significant increase in the no-load losses of the machine compared with normal sine-wave operation. However, as shown in the next section, the full-load efficiency is usually not reduced excessively.

The approximate equivalent circuit of Fig. 5.2(b) is similar to that used for normal sine-wave calculations on a locked-rotor induction motor, when the motor current is also limited by the leakage reactance $(X_1 + X_2)$. The standstill or starting behaviour of the induction motor on a sine-wave supply is, therefore, a measure of its harmonic performance. If the motor draws a large starting current, it will also draw large harmonic currents on non-sinusoidal supplies. The leakage reactance of a reluctance or synchronous motor also determines its harmonic current flow. If the motor has a very low leakage reactance, it should be used with caution on non-sinusoidal supplies, since excessive harmonic currents may flow and overheat the motor.

If V_k denotes the kth harmonic component of the supply voltage, the corresponding stator current harmonic is $I_k = V_k/Z_k$, where Z_k is the kth harmonic input impedance. For positive- and negative-sequence harmonics, the approximate equivalent circuit of Fig. 5.2(b) is valid, and $Z_k = k(X_1 + X_2)$.

Thus,

$$I_k = \frac{V_k}{k(X_1 + X_2)}. \tag{5.13}$$

For zero-sequence harmonics, $Z_k = kX_0$, and

$$I_k = \frac{V_k}{kX_0}. \tag{5.14}$$

These formulae permit rapid evaluation of the harmonic currents due to a non-sinusoidal voltage waveform whose harmonic content is known. Usually there are no zero-sequence harmonics and no even-numbered harmonics, and hence the total r.m.s. harmonic current is given by

$$I_{\text{har}} = \sqrt{[I_5^2 + I_7^2 + I_{11}^2 + I_{13}^2 + \cdots + I_k^2 + \cdots]} = \sqrt{\left[\sum_5^{\infty} I_k^2\right]}. \tag{5.15}$$

If I_1 is the fundamental r.m.s. current of the motor, the total r.m.s. stator current, including the fundamental, is

$$I_{\text{r.m.s.}} = \sqrt{[I_1^2 + I_5^2 + I_7^2 + I_{11}^2 + I_{13}^2 + \cdots + I_k^2 + \cdots]} = \sqrt{[I_1^2 + I_{\text{har}}^2]}. \tag{5.16}$$

For a given voltage waveform, the relative harmonic content of the stator current is closely related to the per-unit reactance of the motor, X_{pu}. This is the leakage reactance at fundamental frequency, expressed as a fraction of the base reactance, $X_{\text{base}} = V_R/I_{\text{FL}}$, where V_R is the rated sine-wave phase voltage, and I_{FL} is the rated full-load current.

Thus,

$$X_{\text{pu}} = \frac{(X_1 + X_2)}{X_{\text{base}}} = (X_1 + X_2)\frac{I_{\text{FL}}}{V_R} = \frac{I_{\text{FL}}}{I_S}\sin\phi_s \tag{5.17}$$

where I_S is the fundamental standstill current of the motor and ϕ_s is the motor power factor angle at standstill.

For the usual six-step and twelve-step voltage waveforms, the magnitude of the harmonic voltage is inversely proportional to the order of the harmonic. Thus $V_k = V_1/k$, and equation (5.13) gives the harmonic current

$$I_k = \frac{V_1}{k^2(X_1 + X_2)}. \tag{5.18}$$

If the fundamental phase voltage, V_1, is equal to the rated sine-wave voltage, V_R, then equation (5.17) gives

$$V_1 = V_R = (X_1 + X_2)\,\frac{I_{FL}}{X_{pu}}$$

and substituting this expression in (5.18) gives

$$I_{kpu} = \frac{I_k}{I_{FL}} = \frac{1}{k^2 X_{pu}} \tag{5.19}$$

where I_{kpu} is the per-unit harmonic current based on the rated full-load current.

Using equations (5.15) and (5.19), the total r.m.s. per-unit harmonic current with a six-step voltage supply is evaluated as $0{\cdot}046/X_{pu}$. Similar calculations on a twelve-step supply give a value of $0{\cdot}0105/X_{pu}$. Thus the harmonic r.m.s. current is inversely proportional to the per-unit reactance. The total r.m.s. stator current at full-load on a per-unit basis is $\sqrt{[1 + (0{\cdot}046/X_{pu})^2]}$ for a six-step voltage waveform and $\sqrt{[1 + (0{\cdot}0105/X_{pu})^2]}$ with a twelve-step supply. In Fig. 5.3, the total r.m.s. stator current is plotted as a function of the

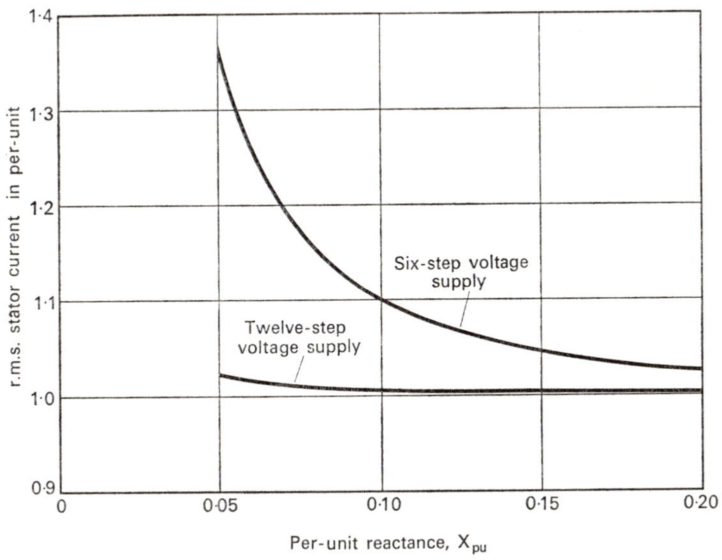

FIG. 5.3. R.M.S. stator current as a function of the per-unit leakage reactance of the motor.

per-unit reactance. The increase in r.m.s. current is almost negligible with a twelve-step supply but the six-step voltage waveform can produce a significant increase, particularly when the per-unit reactance is small. This occurs principally with reluctance motors which may have a per-unit reactance as low as 0·05, resulting in a 35 per cent increase in the r.m.s. full-load current. The polyphase induction motor usually has a per-unit reactance in the range 0·1 to 0·2, and the total r.m.s. current at full-load on a six-step voltage supply is from 2 to 10 per cent greater than the fundamental current.

Figure 5.4 shows a typical stator phase current waveform for an a.c. motor with a six-step voltage supply. This waveform was calculated for a per-unit reactance of 0·1, assuming

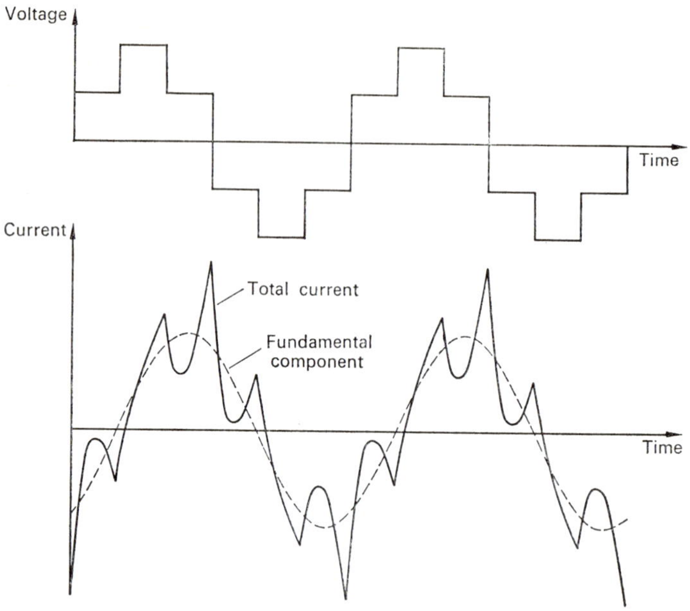

FIG. 5.4. Stator voltage and current waveforms for an a.c. motor with a six-step voltage supply.

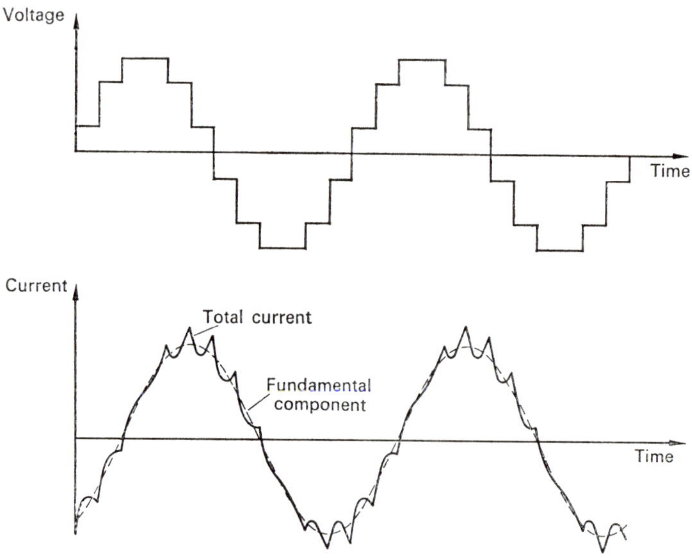

FIG. 5.5. Stator voltage and current waveforms for an a.c. motor with a twelve-step voltage supply.

that the fundamental current lags the fundamental voltage by 60°. This fundamental phase angle is determined by the loading conditions and, in the present case, it corresponds to a fundamental power factor of 0·5. The corresponding current waveform with a twelve-step supply is shown in Fig. 5.5. The harmonic distortion not only increases the r.m.s. value of the stator current, but also produces large current peaks which increase the commutating duty imposed on the static inverter (see Chapter 7). In Fig. 5.6 the ratio of the peak inverter current to the peak fundamental full-load current is plotted as a function of the per-unit reactance. These characteristics were derived theoretically, assuming a fundamental power factor or displacement factor of 0·5. With lower displacement factors, the peak inverter currents are somewhat larger. For an a.c. motor operating on a six-step supply with a fundamental power factor of 0·4, the peak currents are not more than 4 per cent greater than the values indicated in Fig. 5.6.

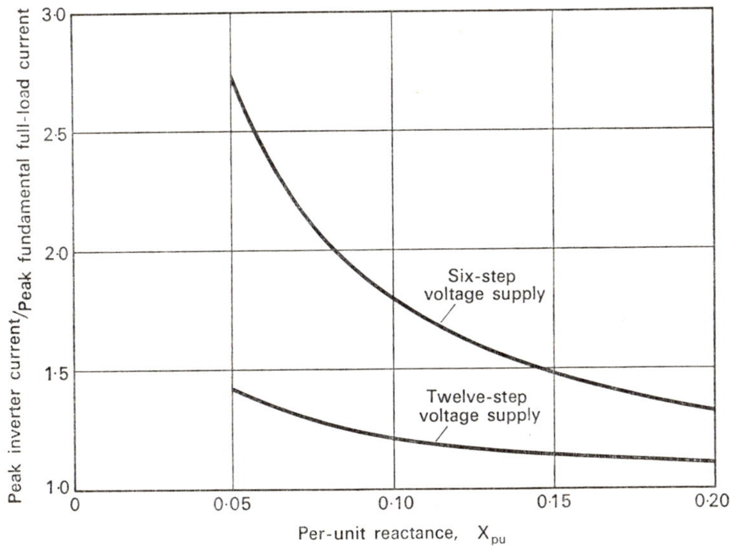

FIG. 5.6. Peak inverter current as a function of the per-unit leakage reactance of the motor.

5.4. MOTOR LOSSES ON NON-SINUSOIDAL SUPPLIES

The additional losses which are present in an a.c. motor with non-sinusoidal excitation are now considered. For comparison with normal sine-wave operation on a 50 Hz a.c. supply, typical numerical values are quoted for operation with six-step voltages having a fundamental frequency of 50 Hz. The harmonic losses are significantly reduced with a twelve-step waveform or a multiple-pulse-width modulated waveform. On the other hand, elementary pulse-width modulation methods with only two or four pulses per half-cycle will have a larger harmonic content and correspondingly greater harmonic losses. In such systems, series inductors may be required in the stator leads in order to reduce the current distortion.

5.4.1. *Stator Copper Loss*

The presence of harmonic currents in the stator winding causes an increased I^2R loss. When skin effect is negligible, the stator copper loss on a non-sinusoidal supply is proportional to the square of the total r.m.s. current. If m_1 is the number of stator phases and

R_1 is the stator resistance per phase, the total stator copper loss is

$$P_1 = m_1 I_{\text{r.m.s.}}^2 R_1 .$$

Substituting for $I_{\text{r.m.s.}}$ from equation (5.16) gives the result:

$$P_1 = m_1 (I_1^2 + I_{\text{har}}^2) R_1 \tag{5.20}$$

where the second term represents the harmonic copper loss. It has been found experimentally that the presence of harmonic currents also increases the fundamental term slightly, due to an increased magnetizing current.[100] This effect is attributed to the increased saturation of the leakage flux paths in the presence of harmonic currents and their associated leakage fluxes.

5.4.2. *Rotor Copper Loss*

The assumption of a constant resistance at the harmonic frequencies is justified for the stator windings of small wire-wound machines, but the rotor a.c. resistance of the induction motor is increased considerably by the skin effect, particularly in the case of deep-bar rotors. In a reluctance or synchronous motor operating on a sine-wave supply, the loss in the rotor cage winding due to the small space harmonic stator m.m.f.s is negligible. When the synchronous machine operates on a non-sinusoidal supply, the time harmonic m.m.f. waves induce rotor harmonic currents as in an induction motor operating near its fundamental synchronous speed. In both synchronous and asynchronous motors, the fifth harmonic m.m.f. rotating backwards and the seventh harmonic rotating forwards will induce rotor currents of six times fundamental frequency, 300 Hz in the case of a 50-Hz supply. Similarly the eleventh and thirteenth harmonics induce rotor currents of twelve times fundamental frequency or 600 Hz. At these harmonic frequencies, the rotor resistance is much greater than the d.c. value. The actual increase depends on the geometrical shape of the conductor cross-section and the rotor slot in which it is placed. Alger has published curves giving the factor by which the resistance is increased.[46] For a rectangular copper conductor of 0·5-inch depth in a rectangular rotor slot, the ratio of a.c. to d.c. resistance at 50 Hz is approximately 1·5, at 300 Hz the ratio is 2·6 and at 600 Hz it is 3·7. At higher frequencies the ratio increases as the square root of the frequency.

Since the rotor resistance is a function of the harmonic frequency, the rotor copper loss is calculated independently for each harmonic. In general, for the kth harmonic

$$P_{2k} = m_1 (I_{2k})^2 R_{2k} \tag{5.21}$$

where I_{2k} is the kth harmonic rotor current, and R_{2k} is the corresponding rotor resistance, corrected for skin effect. The total harmonic copper loss is obtained as a summation of the harmonic contributions. In many induction motors, the additional rotor copper loss due to harmonic currents is the principal cause of reduced efficiency on non-sinusoidal supplies.

5.4.3. *Harmonic Core Loss*

The core loss in the machine is also increased by the presence of harmonics in the supply voltage. As already explained, a time harmonic m.m.f. wave is established in the air gap by each stator current harmonic. These time harmonic waves of m.m.f. have the same number of poles as the fundamental field, but rotate forwards or backwards at a multiple of the fundamental speed. The resultant air-gap m.m.f. at any point is due to the combined effect of the fundamental and time harmonic waves of m.m.f. A detailed analysis shows that the peak of the rotating flux density wave is not constant around the air gap but may be greater or less than the fundamental value.[100] With a three-phase, six-step voltage waveform, the peak flux density in the air gap is approximately 10 per cent greater than the fundamental value, but the overall increase in core loss due to time harmonic main fluxes is a negligible part of the total iron loss. The core loss due to space harmonic fluxes is also negligible, but the end-leakage and skew-leakage fluxes, which normally contribute to the stray-load loss, may produce an appreciable core loss at the harmonic frequencies. Consequently, these effects must be taken into consideration when operating on non-sinusoidal supplies.[101] The end-leakage loss is the eddy-current loss in the end laminations due to leakage flux which enters the laminations in an axial direction. The end-leakage effect is present in both stator and rotor windings, and the end loss is present in both members. The skew-leakage effect is only present in squirrel-cage motors in which the rotor slots are skewed with respect to the stator slots. This construction results in a variable phase difference along the core length between the stator and rotor m.m.f.s. If the m.m.f.s of the stator and rotor conductors balance one another at the centre of the core, there is a resultant radial air-gap m.m.f. as one moves axially from the core centre. This skew m.m.f. is greatest at the core ends, and establishes a skew-leakage flux which produces a core loss in the stator and rotor core and teeth.

5.4.4. *Motor Efficiency*

The magnitude of the harmonic losses obviously depends on the harmonic content of the applied voltage. Large harmonic voltages cause increased machine losses and reduced efficiency. However, most static inverters do not generate harmonics of lower order than the fifth and the high-order harmonics have small amplitudes. For such waveforms, the reduction in motor efficiency is not serious. The various machine losses have been evaluated, and the calculations verified experimentally in the case of an induction motor fed with three-phase, 50-Hz, six-step voltages having a fundamental component equal to the rated sine-wave voltage of the machine.[102,103] For a typical medium horsepower induction motor, the full-load r.m.s. current is approximately 4 per cent greater than the fundamental value. If the skin effect is neglected, the motor copper loss is proportional to the square of the total r.m.s. current, and the harmonic copper loss is therefore 8 per cent of the fundamental loss. Allowing an average threefold increase in rotor resistance due to skin effect, the harmonic copper loss in the motor can be conservatively estimated at 24 per cent of the fundamental loss. If the copper loss constitutes 50 per cent of the total machine losses, this causes an increase of 12 per cent in the overall machine losses.

The increase in core loss is less predictable, as it is influenced by the machine construction and the magnetic materials used. If the high-order harmonic content of the stator voltage waveform is relatively low, as in the six-step wave, the harmonic core loss should not exceed 10 per cent. If the iron loss and stray-load loss constitute approximately 40 per cent of the machine losses, then the harmonic contribution to the total machine losses is only 4 per cent. Friction and windage losses are unaffected and, consequently, the overall increase in motor losses is less than 20 per cent. If the motor has a normal efficiency of 90 per cent on a 50-Hz sinusoidal supply, the presence of harmonics increases the machine losses from 10 per cent to 12 per cent, and the motor efficiency falls by only 2 per cent.

If the harmonic content of the applied voltage waveform is significantly greater than that of the six-step wave, the harmonic motor losses can be increased considerably, and may be greater than the fundamental losses.(100) Even with a six-step voltage supply, a reluctance motor of low leakage reactance may draw very large harmonic currents, causing a drop in motor efficiency of 5 per cent or more. In such cases it may be necessary to use a twelve-step inverter for satisfactory operation. In the induction motor or reluctance motor, the harmonic currents and losses are practically independent of load, and the time-harmonic losses can be determined by comparing no-load losses on sinusoidal and non-sinusoidal supplies.

5.5. HARMONIC TORQUES

The presence of time harmonic m.m.f. waves in the air gap results in additional harmonic torques on the rotor. These torques are of two types: (a) steady harmonic torques and (b) pulsating harmonic torques.

5.5.1. *Steady Harmonic Torques*

Constant or steady torques are developed by the reaction of harmonic air-gap fluxes with harmonic rotor currents of the same order. However, these steady harmonic torques are a very small fraction of the rated torque, and have negligible effect on the motor operation.(64,104) This may be verified by calculating the torque contribution from the harmonic equivalent circuit, just as the fundamental torque is derived from the fundamental equivalent circuit (see Appendix A).

From the kth harmonic equivalent circuit of Fig. 5.1(b),

$$T_k = \frac{pm_1}{2\pi k f_1}(I_{2k})^2\left(\frac{R_{2k}}{s_k}\right)$$

where T_k is the kth harmonic torque.

For normal full-load operation near synchronous speed, the fundamental slip s_1 is small, and substituting in equation (5.12) gives the harmonic slip as

$$s_k = \frac{k \mp 1}{k}.$$

Hence,

$$T_k = \pm\frac{pm_1}{2\pi f_1}(I_{2k})^2\frac{R_{2k}}{(k \mp 1)} \tag{5.22}$$

where a forward harmonic torque is positive, and a backward torque is negative. The approximate rotor currents from Fig. 5.2(b) may be used in equation (5.22), but the rotor resistance R_{2k} must be corrected for skin effect at the harmonic frequency. The resultant electromagnetic torque is obtained by an algebraic summation of the fundamental and harmonic torques. These calculations were performed by the author for a two-horsepower induction motor operating on a 50-Hz six-step supply, and the fifth harmonic torque was found to be only 0·125 per cent of the fundamental torque. This result was obtained assuming a threefold increase in rotor resistance due to skin effect at the harmonic frequency. Thus, the fifth harmonic torque is negligible, and for higher-order harmonics the torque contribution is even less. In addition, the small counter-torque due to the negative-sequence fifth harmonic is opposed by the forward torque of the positive-sequence seventh harmonic. The combined effect of the fifth and seventh harmonics is, therefore, to produce a very small negative torque opposing the fundamental motor torque. This also applies to the eleventh and thirteenth harmonics, and the overall effect of the supply harmonics is a very slight reduction in the steady torque developed by the motor.

5.5.2. *Pulsating Harmonic Torques*

Pulsating torques with a mean value of zero are produced by the reaction of harmonic rotating fluxes with harmonic rotor currents of a different order. The principal pulsating torques arise from the interaction between the fundamental rotating flux and the harmonic rotor currents. The fifth harmonic stator currents, for example, form a negative-sequence system and produce a space fundamental m.m.f. wave which rotates at five times the fundamental synchronous speed in the opposite direction to the fundamental field. The rotor currents induced by this time harmonic field react with the fundamental rotating field to produce a pulsating torque at six times fundamental frequency, since the relative speed of the rotor current wave and the fundamental m.m.f. wave is six times synchronous speed.

The seventh harmonic stator currents also produce a pulsating torque at six times fundamental frequency. The seventh harmonic has positive phase sequence and therefore produces a time harmonic field rotating at seven times synchronous speed in the same direction as the fundamental field. The relative speed of the main air-gap field and the harmonic current wave in the rotor is again six times fundamental frequency, and the two pulsating torques at six times fundamental frequency combine to produce a fluctuation in the electromagnetic torque developed by the motor. Similarly, the eleventh and thirteenth harmonics produce a twelfth harmonic pulsating torque, but the sixth harmonic component predominates in the case of a six-step supply. If a pulse-width modulated inverter with a low switching frequency is employed, severe torque pulsations can occur at the switching frequency.[105] However, in general, there is no alteration in the steady-state torque of the motor since the pulsating torques have zero average value, but their presence causes the angular velocity of the rotor to vary during a revolution. At very low speeds, the motor rotation takes place in a series of jerks or steps, and the presence of the pulsating torques may set a lower limit to the useful speed range of the motor. The point at which the speed pulsations become objectionable in a variable-speed drive depends on the inertia of the rotating system and the nature of the application. The torque pulsation can be reduced by operating with an improved voltage waveform, such as a twelve-step wave or a pulse-width modulated wave with a high switching frequency. Increasing the number of stator phases also reduces the amplitude of the torque pulsation.[106-7]

Since the fundamental air-gap flux and the harmonic rotor currents are practically independent of loading conditions, the pulsating torque amplitude is also independent of the load. For induction motors operating on a three-phase, six-step supply, the peak pulsating torque is typically about 10 per cent of the full-load torque, and is somewhat greater in the case of reluctance motors. The instantaneous electromagnetic torque variation during one cycle of the supply frequency is shown in Fig. 5.7 for a two-horsepower induction motor

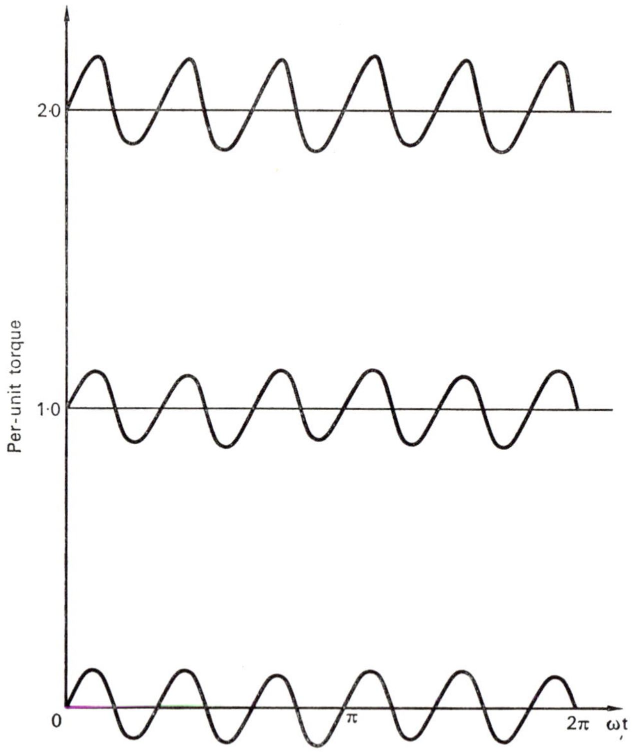

FIG. 5.7. Induction motor torque pulsations at no-load, at rated torque and at twice rated torque, with a six-step voltage supply.

operating on a six-step voltage supply. The stator frequency is 50 Hz and the fundamental voltage is equal to the rated stator voltage of the machine. These torque waveforms were obtained by the author using a digital simulation of the induction motor. Similar torque oscillograms have been obtained by other authors using a linearized set of system equations which assume constant rotor speed and zero inverter regulation. However, a hybrid computer simulation has shown that the pulsating torque amplitude may be considerably accentuated by the voltage fluctuations produced in a practical system with finite inverter regulation.[108] The speed fluctuations associated with a finite load inertia also reinforce the torque oscillations, and it has been found that the torque pulsations at low supply frequencies may be two or three times the theoretical values predicted by an analysis which assumes constant terminal voltage and speed.

5.6. MOTOR INSTABILITY

When a.c. motors are operated on variable-frequency supplies, system instability may occur for certain critical frequency ranges and loading conditions. Machines which are perfectly stable on an infinite system may become unstable with an inverter supply, and machines which are stable when operated individually may become unstable when several of the motors are operated simultaneously as a group drive. The sources of this instability have been investigated, and the two causes which have been discovered are: (a) inherent low-frequency instability in the motor and (b) instability due to interaction between the motor and inverter.

Inherent motor instability is particularly troublesome in the case of the reluctance motor. Even when operated from a balanced sinusoidal voltage source of negligible internal impedance, the reluctance machine can exhibit instability at supply frequencies in the region of 5 Hz to 20 Hz. This instability manifests itself as a sustained oscillation or hunting of the rotor speed about the synchronous value. The resulting variation in torque angle produces corresponding pulsations in the output torque and power. If the rotor oscillations become excessively large, the pull-out torque is exceeded and the machine loses synchronism. Alternatively, the commutating capability of the inverter may be exceeded, and it trips out.

The inherent instability of the reluctance motor has been studied theoretically and experimentally[109-10] and it has been shown that machine stability is improved by reducing the stator and rotor resistances and also by reducing the direct-axis/quadrature-axis reactance ratio. An increase in stator and rotor leakage reactances also reduces the tendency of the machine to oscillate, and machine stability generally improves with an increase in the pole number. Lawrenson and Bowes[110] have shown that the direct-axis/quadrature-axis resistance ratio of the rotor circuits has a pronounced effect on machine stability. The optimum value is about 0·5 and, when operating near this optimum ratio, the beneficial effects on machine stability of a reduction in stator resistance or an increase in leakage reactance are considerably enhanced.

Inherent instability in the d.c.-excited synchronous motor and the induction motor has also been investigated.[111-12] A well-designed d.c.-excited synchronous motor should not exhibit instability at low speeds but it may become unstable in certain cases. The transient response of the induction motor becomes more oscillatory as the supply frequency is reduced (see Section 6.4) but the normal machine does not usually become unstable on an infinite system. However, small motors with a low inertia constant may be unstable. Induction motor stability may be improved by reducing the magnetizing reactance and increasing the stator and rotor resistances.

Instability due to interaction between the motor and converter can occur with synchronous, reluctance or induction motors. It arises when the converter which supplies the variable-speed motor has a finite source impedance. The impedance may be introduced by a transformer or by the filter which smooths the d.c. link supply in a static inverter drive. System instability usually occurs at frequencies below 25 Hz when an interchange of energy takes place between the motor inertia and the filter inductance and capacitance.[113]

In a variable-speed drive, the unstable region of operation is normally confined to a certain torque and frequency range. A torque-frequency diagram can be prepared as in Fig. 5.8, in which the unstable zone is enclosed by the contour. The critical contour can be predicted by using the method of D-decomposition[110] or by applying the root-locus tech-

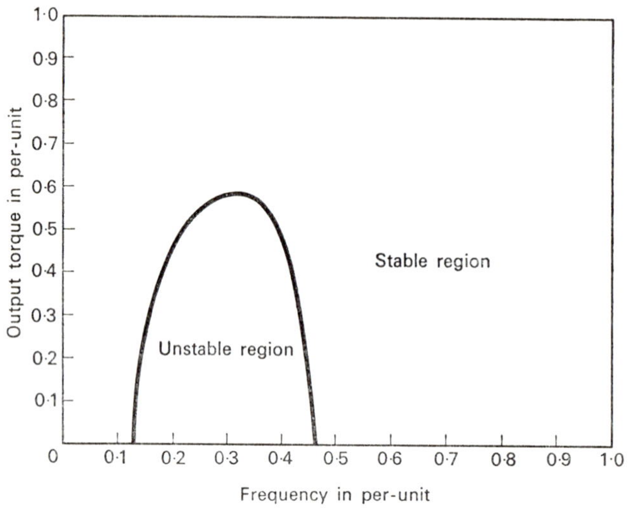

FIG. 5.8. Stability boundary for a converter-fed a.c. motor drive.

nique or the Nyquist stability criterion to a linearized set of machine equations which are valid for small frequency excursions about a fixed base frequency. It is found that the size of the unstable region is affected by the system inertia, the load damping and the electrical parameters of the motor and the supply source. The harmonic content of the stator voltage waveform may slightly affect the system stability, particularly if the inertia is small. However, system stability can usually be assured over the desired range of torque and speed by proper co-ordination of motor and inverter. Altering the value of the filter capacitance sometimes provides a convenient practical method of eliminating instability.

Appropriate modification of the machine parameters will usually eliminate instability in a static variable-frequency drive. Thus, as explained above, reluctance motor stability may be improved by increasing the leakage reactances or by introducing external inductors in the stator leads. This is readily achieved, but the synchronous performance of the motor is adversely affected and the pull-out torque is reduced. Special induction motor designs may also be used to ensure stability over the working range. However, the provision of special machine designs may be uneconomic, and the steady-state performance and efficiency may not be as good as in a normal motor. Alternative techniques have therefore been developed in which standard a.c. motors are employed, and rotor oscillations are damped by means of closed-loop feedback methods. A reluctance motor can be stabilized by altering the amplitude of the stator voltage in accordance with the fluctuations in rotor speed.[114] An inverter-fed induction motor has been stabilized by controlling the inverter frequency using the motor e.m.f.[35] or a derivative of the d.c. link current.[115] Alternative feedback methods have also given satisfactory stabilization of static variable-frequency drives.

5.7. SIX-STEP CURRENT OPERATION

In this chapter, attention has been concentrated on the harmonic behaviour of a.c. motors which are fed from a non-sinusoidal voltage source. However, pulse-width modulated inverters can be modified to deliver a constant-current supply to an a.c. motor.[19] Using a controlled-current circuit of this type, a six-step current waveform can be delivered to the motor instead of the more usual six-step voltage waveform. A.C. motor performance on a six-step current supply is similar to that on a six-step voltage supply, but the harmonic copper loss is slightly larger and the pulsating harmonic torques vary in magnitude with the current. In addition, instability problems are somewhat accentuated with constant-current operation.

5.8. METHODS OF ANALYSIS

A.C. motor performance on non-sinusoidal supplies has been investigated in this chapter by performing a Fourier analysis of the supply waveform and then obtaining the motor response as a summation of the responses for the fundamental component and the various harmonics. This is a simple method of steady-state analysis and is very suitable for a study of the increased machine losses on non-sinusoidal supplies. Several more sophisticated techniques have been suggested for analysing and predicting the performance of inverter-fed induction motors and reluctance motors. Analogue computer simulations of the a.c. motor and the static inverter have been devised.[116–18] Induction motor performance has been studied using the instantaneous symmetrical component transformation to simplify the system equations.[119–21] An analysis based on linear two-axis machine theory has also been performed[122–3], and the method of multiple reference frames has been described.[124–5]

CHAPTER 6

Variable-frequency Operation of Induction Motors

6.1. INTRODUCTION

The most versatile and reliable variable-speed drive consists of a cage-rotor induction motor which is speed-controlled by variation of the stator frequency. In this chapter, the operation of the induction motor on a variable-frequency supply is examined. It has been shown in Chapter 5 that the presence of harmonics in the supply voltage usually has only a minor influence on the motor performance. The present analysis, therefore, assumes balanced sinusoidal supply voltages.

The theoretical characteristics in this chapter were computed for a 2-horsepower, wound-rotor induction motor with short-circuited slip-rings. The computed characteristics were verified experimentally by measurements on the actual machine. Some of the experimental low-speed characteristics were obtained using the static inverter, since it is the most convenient source of low-frequency power. Where necessary, corrections were made for the presence of harmonics.

6.2. STEADY-STATE PERFORMANCE AT CONSTANT VOLTS/HZ

The steady-state performance of the induction motor is readily analysed by means of the fundamental equivalent circuit of Fig. 5.1 (a). For normal sine-wave operation, the skin effect is usually neglected, and hence the resistances are independent of frequency, while the reactances are proportional to frequency. The rotating flux wave in the air gap induces a counter e.m.f. $\mathbf{E}_1$ in the stator winding. This e.m.f. is less than the applied voltage $\mathbf{V}_1$ due to the voltage drop in $(R_1 + jX_1)$, the stator leakage impedance. Since the presence of space harmonic m.m.f. waves is ignored, the rotating flux wave has a sinusoidal space distribution, and the flux linking each stator turn has a sinusoidal time variation. If Φ denotes the flux per pole of the rotating field, the instantaneous flux linking a full-span stator turn is

$$\phi = \Phi \sin \omega_1 t$$

where $\omega_1 = 2\pi f_1$, the angular frequency of the supply voltage. The induced e.m.f. per turn is therefore

$$e_1 = d\phi/dt = \omega_1 \Phi \cos \omega_1 t$$

and the r.m.s. phase e.m.f. is given by

$$E_1 = \omega_1 \Phi k_w N_1/\sqrt{2} = 4{\cdot}44 k_w f_1 N_1 \Phi \tag{6.1}$$

where N_1 is the number of series turns per phase and k_w is the winding factor. If the winding factor is unity, the usual transformer e.m.f. equation is obtained and, hence, for a motor or transformer, Φ is proportional to E_1/f_1, as already deduced in Section 1.3.

For effective utilization, the air-gap flux of the induction motor must be sustained at all frequencies. A constant air-gap flux is obtained when the ratio E_1/f_1 is constant, but if the stator leakage impedance is small, then V_1 and E_1 are approximately equal. Consequently, the air-gap flux is nearly constant when the ratio V_1/f_1 has a fixed value. This is the constant volts/Hz mode of operation which is commonly used in simple open-loop systems. The linear output voltage-frequency characteristic is provided by the static inverter or cyclo-converter, using the voltage control techniques already described. Unfortunately, the motor performance deteriorates at low frequencies when the air-gap flux decreases. This is shown in the following section.

6.2.1. *Torque Characteristics*

From the equivalent circuit of Fig. 5.1(a), the following phasor equations may be obtained:

$$\mathbf{V}_1 = (R_1 + jX_1)\,\mathbf{I}_1 + \left(\frac{R_2}{s} + jX_2\right)\mathbf{I}_2, \tag{6.2}$$

$$jX_m(\mathbf{I}_1 - \mathbf{I}_2) = \left(\frac{R_2}{s} + jX_2\right)\mathbf{I}_2. \tag{6.3}$$

By definition, the rotor slip is given by

$$s = f_2/f_1 \tag{6.4}$$

where f_1 is the variable stator frequency, and f_2 is the corresponding rotor frequency.

In Appendix A, the motor torque is derived as

$$T = \frac{pm_1}{2\pi f_1}(I_2)^2\,\frac{R_2}{s} \tag{6.5}$$

where m_1 is the number of stator phases, and p is the number of pole-pairs.

On combining equations (6.2), (6.3), (6.4) and (6.5), the torque can be expressed in terms of the applied voltage V_1 and the frequencies f_1 and f_2. The resulting expression is

$$T = \frac{pm_1}{2\pi}\left[\frac{V_1}{f_1}\right]^2 \frac{f_2 X_m^2/R_2}{\left[R_1 + \dfrac{f_2}{f_1 R_2}(X_m^2 - X_{11}X_{22})\right]^2 + \left[X_{11} + \dfrac{f_2 R_1 X_{22}}{f_1 R_2}\right]^2} \tag{6.6}$$

where $X_{11} = X_1 + X_m$, the total stator reactance at the supply frequency f_1, and $X_{22} = X_2 + X_m$, the total rotor reactance at the same frequency.

For fixed volts/Hz operation, the quantity V_1/f_1 in equation (6.6) is constant, and the torque characteristics can be evaluated when the machine parameters are known. In Fig. 6.1 the torque is plotted against rotor frequency, f_2, for a number of stator frequencies. The torque is again expressed in per-unit form, using the rated torque of the motor as a base value. As explained in Section 6.1, the machine parameters used in the calculations were measured on a 2-h.p. motor, but similar results are obtained for larger machines.(24,126) These results show that the induction motor torque decreases rapidly as the stator frequency is reduced. This is caused by the reduction in air-gap flux at low frequency due to the increased influence of the stator resistance. The stator resistance voltage drop at rated current has the same magnitude at all frequencies, and it therefore constitutes a considerably higher fraction of the supply voltage at low frequency than at rated frequency, when it may often be neglected.

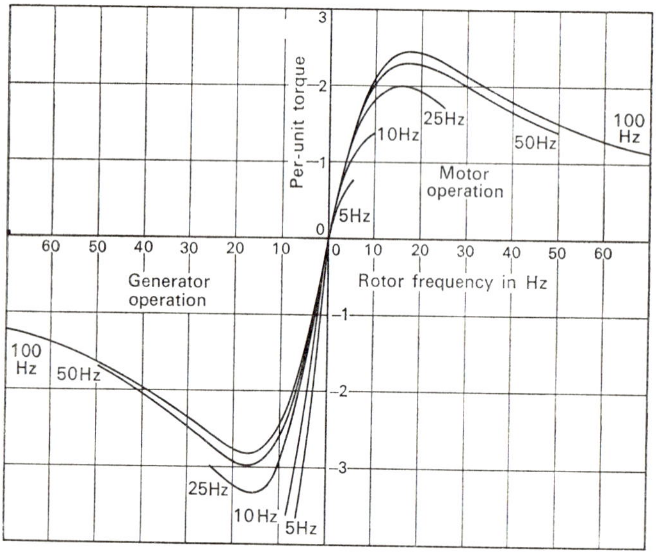

FIG. 6.1. Torque characteristics for variable-frequency operation of the induction motor at constant volts/Hz.

The torque characteristics may be extended into the induction generator region as in Fig. 6.1. In this region, the machine operates with a reversed power flow and reversed stator voltage drop, resulting in an increased e.m.f. E_1, and increased air-gap flux. Consequently, large generator torques are produced, particularly at low frequencies. These braking torques may cause mechanical damage, unless current limiting is employed. However, the characteristics of Fig. 6.1 are based on the linear equivalent circuit of Fig. 5.1(a), in which magnetic saturation is neglected. In practice, saturation effects will cause the braking torques in the generator region to be less than the theoretical values indicated in Fig. 6.1.

In Fig. 6.2 the breakdown and starting torques are plotted against supply frequency for the 2-h.p. motor. The breakdown torque is determined by maximizing the torque expression of equation (6.6), and the starting torque is evaluated by putting f_1 equal to f_2 in equation (6.6). The serious reduction in both these torques at frequencies less than 10 Hz is evident from Fig. 6.2. To improve the low-frequency characteristics, the terminal voltage should be increased considerably above its frequency-proportional value. This is not possible when synchronous alternators are used, but is readily provided by means of static converters which permit independent adjustment of output voltage and frequency.

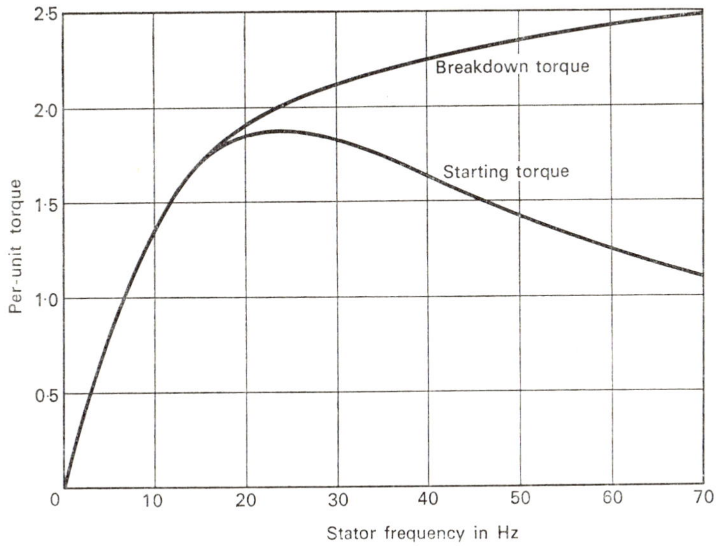

FIG. 6.2. Starting and breakdown torques for variable-frequency operation of the induction motor at constant volts/Hz.

6.2.2. *Stator Current Locus*

The stator current locus at constant volts/Hz can be derived from the equivalent circuit of Fig. 5.1(a), using the method of complex inversion. This is treated in detail by Nürnberg,[127] but the procedure can be summarized briefly as follows: the stator input impedance, $\mathbf{Z}_1$, is a function of the rotor slip, s. The stator current $\mathbf{I}_1 = \mathbf{V}_1/\mathbf{Z}_1$, where $\mathbf{V}_1$ is the applied phase voltage. Consequently, if the locus of $\mathbf{Z}_1$ is determined and then inverted with respect to the origin, the current locus is obtained. In order to place the eventual current diagram in its usual position, the positive resistance axis is drawn vertically upwards, and the positive reactance axis is drawn horizontally to the right.

From the equivalent circuit, the input impedance is expressed by

$$\mathbf{Z}_1 = R_1 + jX_1 + \frac{jX_m\,[(R_2/s) + jX_2]}{(R_2/s) + j\,(X_m + X_2)} = R_1 + jX_{11} + \frac{X_m^2}{(R_2/s) + jX_{22}}.$$

As the slip varies from minus to plus infinity, the input impedance traces a circular locus. Two points on the circle are readily determined: at synchronism, $s = 0$ and $\mathbf{Z}_1 = R_1 + jX_{11}$, while at $s = \infty$, $\mathbf{Z}_1 = R_1 + jX_{11} - jX_m^2/X_{22}$. These two values define the end-points A and B of the horizontal diameter of the impedance circle in Fig. 6.3. The diametral length is X_m^2/X_{22} and is independent of rotor resistance, R_2. The centre point M of the circle is at $(R_1 + jX_{11} - \frac{1}{2}jX_m^2/X_{22})$ and its position is also independent of R_2. A circle transforms into another circle after inversion with respect to a point outside it and, consequently, the impedance circle transforms into an admittance circle after inversion with respect to the origin. If the scales are suitably chosen, the admittance circle coincides with the impedance circle. The stator current is proportional to the admittance, and hence, the admittance circle also represents the current locus. The current scale is obtained by multiplying the admittance scale by the phase voltage.

When the induction motor operates on a variable-frequency supply, the inductive reactances vary linearly with frequency, but the winding resistances remain constant. Under constant volts/Hz conditions, the impedance locus at different supply frequencies can be obtained by assuming that the supply voltage and frequency remain constant, while the

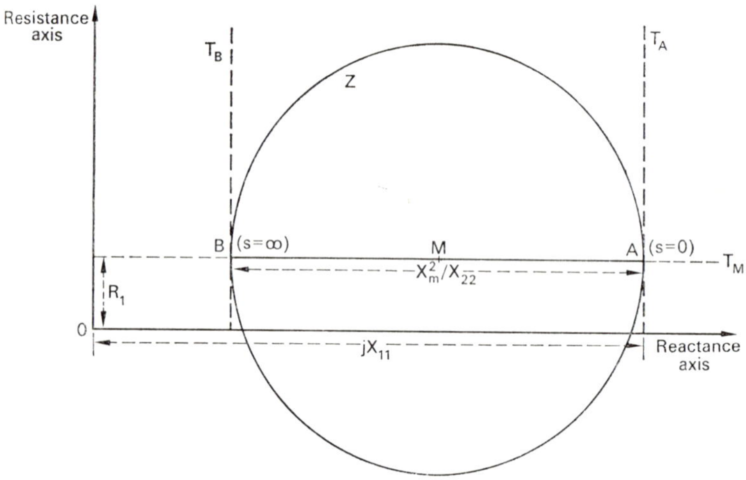

FIG. 6.3. Input impedance locus for the induction motor.

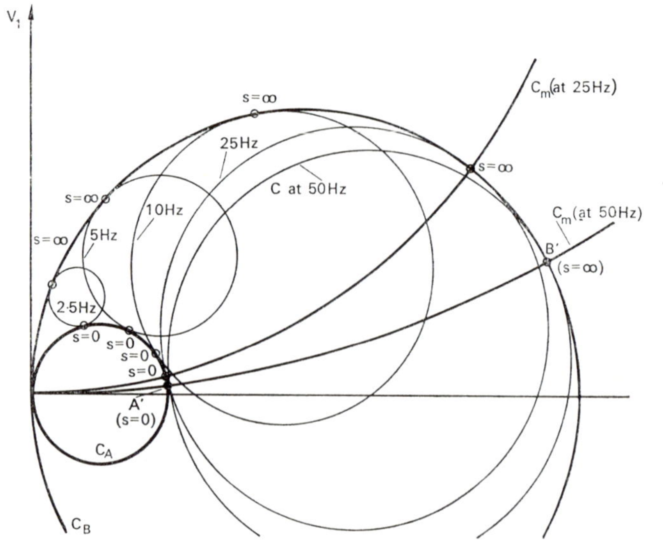

FIG. 6.4. Derivation of induction motor current locus for variable-frequency operation at constant volts/Hz.

resistances R_1 and R_2 vary inversely with frequency. As the supply frequency is reduced, the size of the impedance circle is unaffected, but the equivalent increase in R_1 causes the impedance circle to rise vertically between the tangents T_A and T_B in Fig. 6.3. After inversion, the tangents transform into the circles C_A and C_B of Fig. 6.4. The horizontal diameter T_M transforms into the circle C_M, intersecting C_A and C_B at A′ and B′. These are the inverse points corresponding to A and B on the impedance locus. The impedance circle Z transforms into the current circle C, touching C_A and C_B at the points A′ and B′.

As the supply frequency is reduced at constant volts/Hz, the circles C_A and C_B are unchanged, since the tangents T_A and T_B are unaffected. However, the impedance circle and its horizontal diameter T_M move vertically upwards, and the circular current locus obtained after inversion decreases in size as it moves into the wedge between circles C_A and C_B. At the higher frequencies, the stator resistance is relatively small compared with the leakage reactance, and Fig. 6.4 shows that the stator current locus does not change appreciably. As the frequency is reduced below 10 Hz, however, the resistance becomes increasingly significant, and the current circle rapidly decreases in size. In practice, magnetic saturation is often significant at low frequencies and this causes distortion of the circular current locus.[128]

6.2.3. *Low-frequency Performance with Increased Volts/Hz*

As explained above, the low-speed performance of the induction motor is unsatisfactory at constant volts/Hz, and a voltage boost is required at low frequencies in order to maintain the air-gap flux. A closed-loop feedback system is possible, in which the air-gap flux is monitored and compared with a fixed reference voltage representing the desired flux value. The resulting error voltage is amplified and used to control the terminal voltage at the value required to maintain constant air-gap flux for all loads and stator frequencies. This method is unduly complicated and costly for many applications. A simple open-loop system which is often used, employs a frequency changer with a fixed output voltage-frequency characteristic. This characteristic is usually linear at higher frequencies, but the output voltage is increased for low-frequency operation. The terminal voltage is independent of the loading conditions and, consequently, at low speeds on light loads the motor may have a high degree of magnetic saturation.

The usual equivalent circuit of Fig. 5.1 (a) can be used to calculate the low-frequency performance with an increased terminal voltage. Despite the magnetic saturation, core losses are negligible because of the low stator frequency. Variations in magnetizing reactance, X_m, due to saturation may have to be taken into account, but this can be done graphically as described by Dixon and Tiley.[129] As the frequency of operation is reduced, the reactances decrease proportionately with frequency but the resistances remain constant. At low frequencies, the voltage drop across the stator resistance is, therefore, relatively large compared with the terminal voltage, and the approximate equivalent circuit in which X_m is moved to the input terminals is not valid.

Some unusual effects are noted when the machine is operated at low frequency with an increased terminal voltage. The stator resistance voltage drop is large compared with the leakage reactance drop, and the phasor diagrams of Fig. 6.5(a) and Fig. 6.5(b) show the

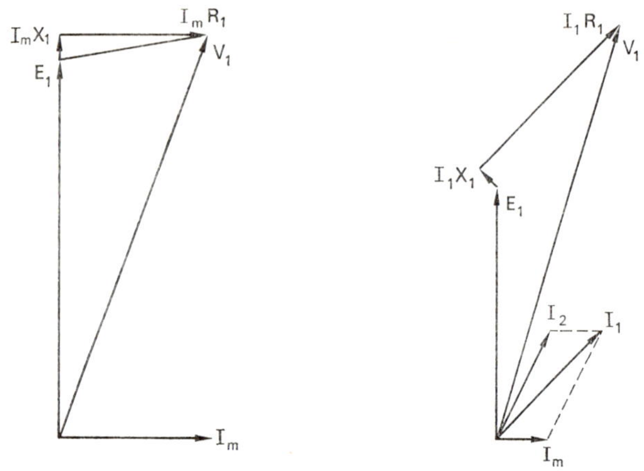

FIG. 6.5. The influence of magnetic saturation on induction motor performance at low frequency: (a) no-load phasor diagram; (b) full-load phasor diagram.

no-load and full-load conditions respectively. The magnetizing current, $\mathbf{I}_m$, lags the air-gap e.m.f., $\mathbf{E}_1$, by 90° and, at no-load, the magnitude of the e.m.f. E_1 is almost equal to the applied voltage, V_1. Thus, the machine is highly saturated and a large magnetizing current flows. This no-load current may equal or even exceed the full-load current.

On load, the rotor current $\mathbf{I}_2$ is more in phase with $\mathbf{E}_1$, and the air-gap e.m.f. is reduced, as shown in Fig. 6.5(b). This causes a large drop in magnetizing current, which therefore decreases rapidly as the load current increases. The reduction in magnetizing current more than offsets the increase in rotor current, and hence the total supply current decreases as

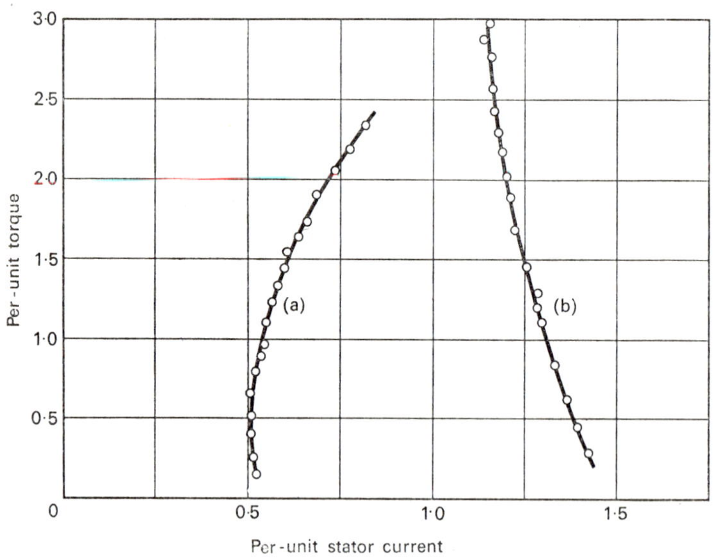

FIG. 6.6. Relationship between induction motor torque and stator current for a supply frequency of 5 Hz: (a) at rated volts/Hz; (b) at 140 per cent of rated volts/Hz.

load is applied. This is illustrated by the characteristics of Fig. 6.6. which were measured at 5 Hz. The relationship between torque and stator current is plotted for the constant volts/Hz condition and also for a terminal voltage which is 40 per cent greater than the frequency-proportional value. The unusual characteristic obtained at the higher voltage is evidently caused by magnetic saturation. This characteristic shows that appreciable torque can be developed at low frequencies provided the terminal voltage is increased sufficiently.

6.3. CONSTANT-FLUX OPERATION

In order to obtain high torque throughout the speed range, the air-gap flux should be maintained constant and not allowed to decrease at low frequencies as a result of increasing resistance effects. This is achieved if the air-gap e.m.f. E_1, rather than the terminal voltage V_1, is varied linearly with frequency.

6.3.1. *Torque Characteristic*

From the equivalent circuit of Fig. 5.1 (a), the rotor current, I_2, is given by the equation

$$I_2 = \frac{E_1}{\sqrt{[(R_2/s)^2 + X_2^2]}}. \tag{6.7}$$

On combining this expression with equations (6.4) and (6.5), the following torque equation is obtained:

$$T = \frac{pm_1}{2\pi}\left[\frac{E_1}{f_1}\right]^2\left[\frac{f_2R_2}{R_2^2 + (2\pi f_2L_2)^2}\right] \tag{6.8}$$

where L_2 is the rotor leakage inductance.

Since the air-gap flux is proportional to (E_1/f_1), the electromagnetic torque is proportional to the square of the air-gap flux at a given rotor frequency, f_2. Consequently, if the air-gap flux is maintained constant under all operating conditions, the induction motor torque is determined solely by the absolute rotor frequency, f_2, and is independent of the supply frequency, f_1. A control scheme in which the air-gap flux is constant and the rotor slip frequency is directly controlled, thus permits a precise adjustment of motor torque at any speed. This form of control is considered in Section 6.7.

By maximizing equation (6.8), an expression is obtained for the breakdown torque with constant air-gap flux. Differentiating with respect to rotor frequency, f_2, and equating to zero, yields the rotor breakdown frequency as

$$f_{2b} = \pm\frac{R_2}{2\pi L_2} \tag{6.9}$$

where the negative sign applies to generator operation.

On substituting this value in equation (6.8), the breakdown torque is obtained as

$$T_b = \pm \frac{pm_1}{2\pi} \left[\frac{E_1}{f_1}\right]^2 \frac{1}{4\pi L_2}. \tag{6.10}$$

Thus, the breakdown torque is proportional to the square of the air-gap flux and inversely proportional to the rotor leakage inductance. The rotor resistance does not affect the breakdown torque, but influences the rotor frequency at which breakdown occurs. On substituting equations (6.9) and (6.10) in the earlier torque expression (6.8), the following universal torque equation is obtained for constant-flux operation:

$$\frac{T}{T_b} = \frac{2}{(f_2/f_{2b}) + (f_{2b}/f_2)}. \tag{6.11}$$

Figure 6.7 shows the theoretical torque characteristic for the 2-h.p. motor, assuming the air-gap flux is maintained constant at the value corresponding to full-load operation at rated voltage and frequency. This characteristic is valid at all stator frequencies, and for

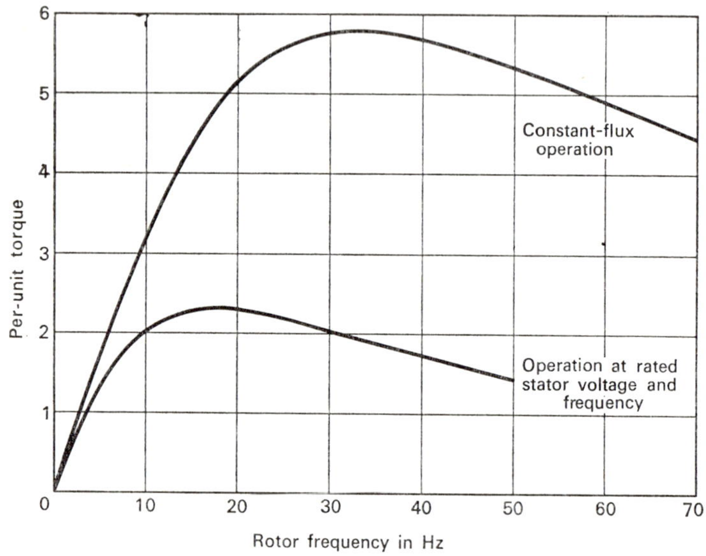

FIG. 6.7. The universal torque characteristic for constant-flux operation of the induction motor compared with the normal torque characteristic at rated voltage and frequency.

motor or generator operation. The torque characteristic for normal fixed-frequency operation at rated voltage and frequency is included in Fig. 6.7 for comparison. Under constant-flux conditions, the available torque is considerably larger and can be obtained over the complete range of supply frequencies, thus eliminating any deterioration in low-speed performance.

6.3.2. *Stator Current Locus*

The large torque available under constant-flux conditions cannot be obtained without correspondingly large machine currents. From the equivalent circuit, the stator current, $\mathbf{I}_1$, is given by

$$\mathbf{I}_1 = \mathbf{I}_m + \mathbf{I}_2 = \frac{\mathbf{E}_1}{jX_m} + \frac{\mathbf{E}_1}{(R_2/s) + jX_2}$$

$$= \left[\frac{\mathbf{E}_1}{f_1}\right] \cdot \left[\frac{R_2 + j(2\pi L_{22}) f_2}{(2\pi L_m)^2 f_2 - (2\pi L_m)(2\pi L_{22}) f_2 + j(2\pi L_m) R_2}\right] \tag{6.12}$$

where L_m is the magnetizing inductance, and $L_{22} = L_2 + L_m$, the total rotor inductance.

Equation (6.12) shows that the stator current, $\mathbf{I}_1$, is independent of the supply frequency, f_1, when the air-gap flux is constant. If the stator current locus is plotted according to equation (6.12), a general circle diagram is obtained which is valid at all stator frequencies. The circular locus has a larger diameter than the normal circle diagram at rated voltage and frequency, indicating that the machine currents are greater.(126) Sustained operation in the high torque region may not be possible unless special cooling methods are provided.

6.3.3. *Stator Voltage*

The terminal voltage, $\mathbf{V}_1$, is obtained by phasor addition of the air-gap e.m.f., $\mathbf{E}_1$, and stator voltage drop, $(R_1 + jX_1)\mathbf{I}_1$. Thus,

$$\mathbf{V}_1 = \mathbf{E}_1 + (R_1 + jX_1)\mathbf{I}_1. \tag{6.13}$$

Under constant-flux conditions, the air-gap e.m.f. E_1 is varied linearly with stator frequency, but the stator voltage drop is determined by the supply current, I_1, and frequency, f_1. The terminal voltage necessary to maintain a constant flux density in the machine is,

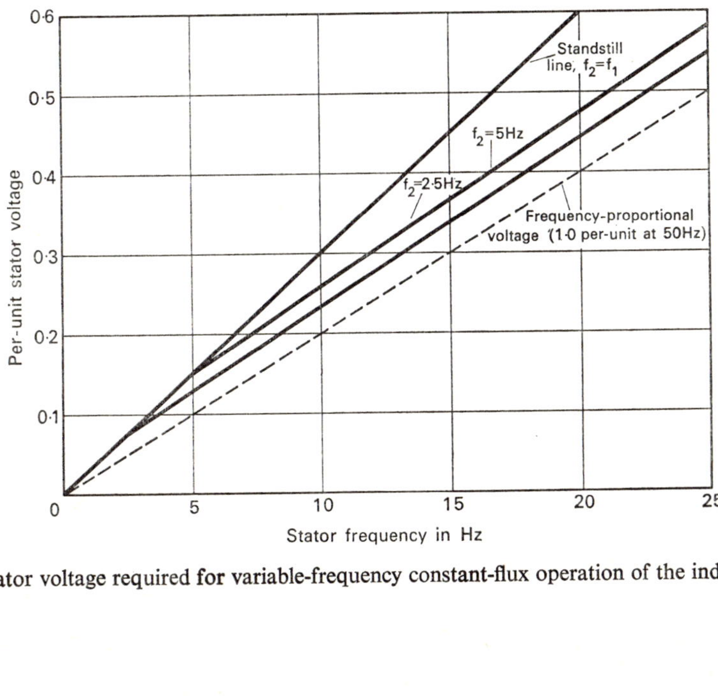

FIG. 6.8. Stator voltage required for variable-frequency constant-flux operation of the induction motor.

therefore, a function of stator frequency and motor load. The loading conditions determine the rotor frequency, f_2, and hence the stator current, $\mathbf{I}_1$, as expressed by equation (6.12). Curves of terminal voltage as a function of stator frequency can be plotted with the rotor frequency as parameter.(126,130) These characteristics were computed using the same machine parameters as before, and are plotted in Fig. 6.8. The increased terminal voltage required by a loaded induction motor at low frequencies is clearly seen.

6.4. CONSTANT-CURRENT OPERATION

The air-gap flux of the machine can be indirectly determined by controlling the stator current, I_1, and rotor slip frequency, f_2. This method of control is examined in more detail in Section 6.6 but, for the moment, it can be noted that since the motor torque is, in fact, determined by the air-gap flux and rotor current, there are advantages in directly controlling the machine current rather than the terminal voltage. There are also benefits arising from the operation of the converter as a constant-current supply. Since the converter current is controlled at a constant value, there are no transient current surges, and a large over-current capacity is unnecessary. This makes possible a more economic inverter design in which the thyristors are fully utilized during normal operation.

The motor torque can be expressed in terms of the stator current and rotor frequency by combining equations (6.3), (6.4) and (6.5). Thus,

$$T = \frac{pm_1}{2\pi}(I_1)^2\left[\frac{(2\pi L_m)^2}{(R_2/f_2) + (f_2/R_2)(2\pi L_{22})^2}\right]. \tag{6.14}$$

This is the theoretical output torque, neglecting saturation, and it is obvious that large torques can be developed at any stator frequency by supplying large stator currents to the machine.

The air-gap flux is proportional to the magnetizing current, I_m, assuming that the magnetizing inductance, L_m, is unaffected by saturation. In order to preserve constant air-gap flux in the motor, the magnetizing current must be held constant. The stator current, I_1, required to achieve this condition is a function of the rotor frequency, f_2, but is independent of the stator frequency, f_1. The exact relationship between I_1 and I_m is

$$I_1 = I_m\sqrt{\left[\frac{1+(f_2)^2(2\pi L_{22}/R_2)^2}{1+(f_2)^2(2\pi L_2/R_2)^2}\right]}. \tag{6.15}$$

When constant-flux operation is required, the prescribed values of stator current and rotor frequency must always be suitably related.

It is readily shown that, for a given stator current, the torque expression in equation (6.14) has a maximum value when the rotor frequency $f_2 = R_2/(2\pi L_{22}) = R_2/[2\pi(L_2 + L_m)]$. At this breakdown frequency, the torque per ampere of the motor has an optimum value. Theoretically, the rotor breakdown frequency has the same value for all stator currents but, in practice, because of magnetic saturation and the resulting variation in machine parameters, the breakdown frequency is a function of the stator current. As the current increases, magnetic saturation causes a reduction in the inductances, and hence the rotor breakdown frequency is displaced to a higher value.[126,131] For optimum performance in a controlled-current system, the rotor slip frequency should also be directly controlled and adjusted to the optimum value depending on the stator current. Due to magnetic saturation, the motor torque is also less than the theoretical value predicted by equation (6.14). In practice, transient torques of six times rated torque can be obtained with stator currents of four or five times the rated value.

6.5. TRANSIENT PERFORMANCE OF THE FREQUENCY-CONTROLLED INDUCTION MOTOR

The development of static frequency converters for speed control applications has led to an increased interest in the transient performance of the induction motor on a variable-frequency supply. A general set of equations which is valid for transient and steady-state conditions can be derived from a simplified model of the induction motor. This model has sinusoidal m.m.f. and flux density distributions and has no magnetic saturation and no hysteresis or eddy-current losses. In spite of these simplifications, the resulting differential equations are non-linear and rather complicated. They can be somewhat simplified by means of the *dq*-axis transformation but, if the speed is variable, the equations remain non-linear, and a general solution is difficult to obtain analytically. Usually, an analogue computer is used to simulate the machine equations,[132–4] or a numerical solution is obtained on a digital computer.[135–7]

Lawrenson and Stephenson[138] have used the digital computer to study the transient performance of a particular induction motor at variable supply frequency. They have shown that the motor speed can be changed rapidly and efficiently by varying the supply frequency uniformly with time at constant volts/Hz. Under these conditions, a run-up from standstill can be achieved more quickly than by the usual step application of power at rated frequency. The machine currents are also smaller, but the optimum rate of change of frequency is dependent on the load inertia. A complete speed reversal from full speed in one direction to full speed in the opposite direction also takes a shorter time and has lower losses when the voltage and frequency are varied uniformly with time. When the usual plugging method is used, the phase sequence of the supply voltage is suddenly reversed, thereby changing the direction of the field rotation. As the motor is brought to rest, the kinetic energy in the rotating masses is dissipated as losses in the rotor circuit, and this may give rise to heating problems. When the voltage and frequency are reduced uniformly, the motor is brought to rest by regenerative braking, since the supply frequency is reduced faster than the rotor speed, and the machine operates with negative slip. The kinetic energy of the rotating masses is therefore returned to the supply and is not dissipated in the rotor.

The induction motor also exhibits oscillatory behaviour which may prove objectionable under variable-frequency conditions of operation. If a lightly loaded induction motor is run-up to normal speed by the sudden application of voltage, it is found that the speed may overshoot the synchronous value and oscillate about synchronism for a few cycles before steady speed conditions are achieved. Similar speed oscillations are obtained when the load torque is suddenly altered, or the supply voltage is changed rapidly. It has been shown that the damping coefficient for these speed oscillations is a function of the supply frequency. As the stator frequency is reduced below the rated value, the speed oscillations become more persistent, and the damping coefficient has a minimum value at some reduced frequency which is dependent on the system inertia. For a normal 50- or 60-Hz motor, the damping is least when the supply frequency lies between 10 and 20 per cent of the rated frequency. The lightly damped oscillations occurring at these reduced frequencies impair the system performance at low speeds. Small induction motors may even become unstable, as mentioned in Section 5.6.

In order to facilitate the transient analysis of the induction motor, the general *dq*-axis equations may be simplified by considering linearization about a steady operating point. This provides a set of linear differential equations which are valid for small deviations from the steady-state condition.(112) The stability of the machine can then be investigated by applying the conventional methods of linear feedback control theory. Rogers(139) has used the root locus method to analyse induction motor transients due to small disturbances in stator voltage and load torque. Pfaff(140) and De Carli *et al.*(141) have also made approximations to the machine equations in order to derive simple models of the induction motor which are valid for small frequency oscillations about a fixed base frequency. An idealized two-coil rotor model of the induction motor has been used to study the transient performance on a constant-current supply.(142–3) Other approximate methods of analysis have been developed, but in the case of complex variable-speed drives, it appears that a computer simulation of the system is necessary at the design stage in order to predict the transient performance and to ensure that system instability does not occur in the operating range.

6.6. CLOSED-LOOP OPERATION

The open-loop operation of an induction motor at variable frequency provides a satisfactory variable-speed drive when the motor is required to operate at steady speeds for long periods. When the drive requirements include rapid acceleration and deceleration, an open-loop system is unsatisfactory, since the supply frequency cannot be varied very quickly, or the rotor breakdown frequency is exceeded. Beyond the breakdown point, the rotor current is large, but the power factor is low and the output torque and efficiency are small. When a fast dynamic response is necessary, closed-loop feedback methods are essential. The slip frequency can then be controlled so that operation always occurs at small slip, thereby yielding high torque at high power factor with low losses. Optimization of the supply conditions for the motor is also possible. This means that the motor voltage and frequency are adjusted to the values which give the required steady-state torque at the specified speed with minimum losses in the machine.(144) In a closed-loop system, the drive characteristics can also be modified by the use of feedback. Thus, the system can be designed to maintain constant torque over a wide speed range down to standstill. Alternatively, the drive can develop a constant-horsepower output which is suitable for traction applications.

6.6.1. *Control Techniques*

As shown in Appendix A, the induction motor torque is proportional to the product of air-gap flux and in-phase rotor current, and hence

$$T = K\Phi I_2 \cos \phi_2.$$

In order to develop a certain motor torque, the machine can be operated with a large air-gap flux and a small rotor current, or a small flux and a large current. The air-gap flux can be increased by raising the stator voltage/frequency ratio, but magnetic saturation ultimately occurs, resulting in excessive core loss and magnetizing current. The rotor current can be increased by raising the rotor slip frequency, f_2, until large currents at low power factor produce excessive copper loss and overheating of the motor. An optimum rotor frequency exists for each loading condition which gives the required torque at the required speed with minimum losses in the machine. The optimum conditions are not very critical, however, and the control circuitry is often simplified by operating the motor with a constant rotor frequency and variable air-gap flux.

In a closed-loop system, the rotor frequency is always restricted to small values below the breakdown point. At small slip, $\cos \phi_2 \simeq 1$, and the general torque equation above is simplified to

$$T = K\Phi I_2. \tag{6.16}$$

The rotor current, in general, is calculated from the equivalent circuit of Fig. 5.1(a) as

$$I_2 = \frac{sE_1}{\sqrt{[R_2^2 + (sX_2)^2]}}.$$

When s is small, $sX_2 \ll R_2$, and

$$I_2 = \frac{sE_1}{R_2} = \frac{f_2}{f_1} \cdot \frac{E_1}{R_2}.$$

Since the air-gap flux, Φ, is proportional to E_1/f_1, the rotor current is proportional to the product of air-gap flux and absolute slip frequency. Hence,

$$I_2 = K'\Phi f_2. \tag{6.17}$$

On combining equations (6.16) and (6.17)

$$T = K''\Phi^2 f_2. \tag{6.18}$$

Thus, with constant air-gap flux and small rotor slip, torque is directly proportional to the absolute rotor frequency. The actual speed has no effect on the shaft torque, apart from a slight reduction with increasing speed due to windage and friction. The torque equation is valid for positive or negative slip frequencies corresponding to motor or generator operation. For motor operation, a predetermined rotor frequency, f_2, is obtained by applying a stator frequency $f_1 = f_n + f_2$, where f_n is the motor rotational frequency ($f_n = np$, where n is the shaft speed in revolutions per second and p is the number of pole-pairs). Negative slip implies generator operation with a supply frequency, f_1, which is less than the rotational frequency by an amount f_2, that is, $f_1 = f_n - f_2$. Motor or generator operation with a predetermined rotor frequency is therefore obtained by closed-loop control of the absolute slip frequency.

The air-gap flux can also be controlled in a closed-loop system. The actual air-gap flux is monitored using a Hall generator or stator search coil, and is adjusted to the desired level by variation of the terminal voltage so that V_1/f_1 has the required value. If f_2 and Φ are constrained to remain constant, then the rotor current, I_2, is also constant, as specified by equation (6.17). A constant air-gap flux demands a fixed magnetizing current, I_m, and hence the stator current, $\mathbf{I}_1 = \mathbf{I}_m + \mathbf{I}_2$, is also fixed. Consequently, a closed-loop system in which f_2 and Φ are controlled at constant values, also has a constant stator current and develops a constant shaft torque at all motor speeds. Alternatively, if the stator current itself is adjusted in a feedback loop and the air-gap flux is also controlled, then the rotor frequency is automatically determined without being directly controlled. As a third possibility, the stator current and rotor frequency can be controlled directly, thus providing indirect control of air-gap flux. The advantages of this method of control have been listed in Section 6.4. Induction motor drives in which the rotor frequency is directly controlled are known as controlled-slip systems.

6.7. THE CONTROLLED-SLIP DRIVE[145-8]

Figure 6.9 shows a block diagram of a speed control scheme using the controlled-slip principle. The static converter is assumed to be capable of regenerative operation. This is true for the cycloconverter circuit and also for d.c. link converters in which the d.c. link supply is obtained from the a.c. network by means of a double-converter connection (Section 3.8). The actual motor speed is measured by means of a tacho-generator, and is com-

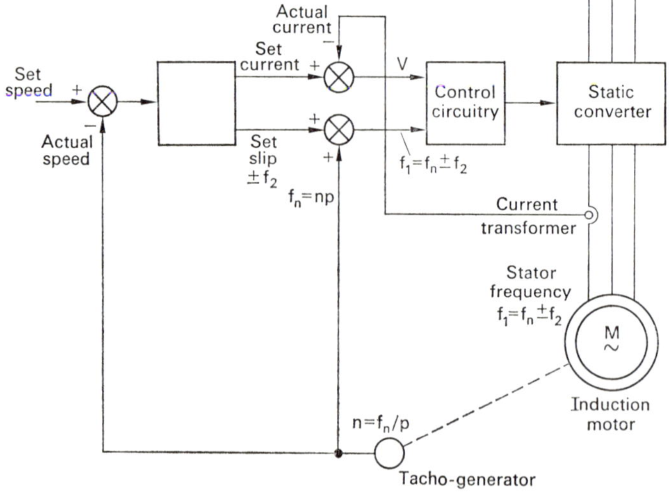

FIG. 6.9. Block diagram of a controlled-slip induction motor drive.

pared with the desired speed. The magnitude and polarity of the speed error are used to determine the stator current and rotor frequency reference values. The actual stator current is monitored by means of current transformers, and the signal is fed to the control circuitry for comparison with the demanded value. If controlled-current operation is not employed, a current limit is usually incorporated, since the maximum current which can be commutated by the inverter has a limiting value which must not be exceeded, even momentarily, or a commutation failure will occur.

The demanded slip frequency is added to, or subtracted from, the rotational frequency measured by the tacho-generator, in order to determine the excitation frequency, f_1, which must be supplied to the motor. If the actual speed is less than the demanded speed, a stator frequency, $f_1 = f_n + f_2$, is applied, and the machine develops a large motor torque which rapidly accelerates the rotor to the desired speed. When the actual speed is greater than the demanded speed, the slip frequency reference value is reversed, and the static converter delivers a stator frequency $f_1 = f_n - f_2$. Under these conditions, the machine operates as an induction generator, returning energy to the a.c. supply through the static converter which also supplies the lagging kVA required by the induction generator. This permits rapid regenerative deceleration of the motor by restoring the kinetic energy of the rotating masses to the a.c. supply. A slip limit is incorporated in the control circuit so that the demanded rotor frequency can never exceed the breakdown value in the motoring or generating regions. Thus, stable operation is possible, close to the breakdown point, giving a large machine torque which improves the dynamic performance of the system for sudden changes in demanded speed.

The addition of the rotational frequency, f_n, and the desired rotor frequency, $\pm f_2$, can be performed mechanically using a differential gear. The differential has one shaft driven directly by the motor at a speed $n = f_n/p$ rev/sec. The differential housing is driven by an auxiliary motor at a speed $0{\cdot}5 f_2/p$, where f_2 is the desired rotor frequency. The output shaft of the differential rotates at a speed $f_1/p = (f_n + f_2)/p$ or $(f_n - f_2)/p$ depending on the relative directions of rotation. This shaft speed determines the excitation frequency, f_1, which is supplied to the stator of the induction motor. The frequency addition may also be performed statically using digital techniques as described by Abraham and Koppelmann.[(148)] A disc on the output shaft of the motor has alternate light and dark segments which rotate past a photo-electric probe, generating a pulse train whose repetition frequency is proportional to the rotational frequency, f_n. A second pulse train whose frequency is proportional to the desired rotor frequency, f_2, is generated and added to the first. The resulting pulses are fed to the control circuitry of the static converter which then generates the required output frequency.

As explained above, a change in the polarity of the slip signal causes a reversal of the machine torque and thereby reverses the direction of power flow. The direction of rotation is determined by the phase sequence of the stator supply. Reversal of the phase sequence is obtained statically by means of the inverter control circuitry, and no switching of power leads is necessary. Independent control of the direction of rotation and the slip polarity means that four-quadrant operation is possible, giving motor or generator operation in either direction of rotation.

In less sophisticated systems, a d.c. tacho-generator on the motor shaft feeds the inverter control circuitry with a control voltage which is adjusted to give a stator frequency equal to the rotational frequency. The tacho-generator voltage is then offset by a fixed voltage which causes the motor to operate with a constant rotor frequency, f_2, for all speeds and loading conditions.

By operating the induction motor in a controlled-slip drive, the overall system characteristics can be adjusted to suit the particular application. Traction applications require large torque at standstill and low speeds, in order to provide rapid acceleration. Low torque is sufficient at higher speeds, and hence a constant-horsepower characteristic is suitable. Constant-horsepower, constant-torque, or constant-speed characteristics can all be obtained from a controlled-slip drive. Torque control is obtained by varying the air-gap flux level or the absolute slip frequency, or both together.

6.7.1. *Constant-torque and Constant-horsepower Operation*

As shown in Section 6.3.1, the induction motor can develop considerable torque at any speed, provided the air-gap flux is constant. The rotor frequency, f_2, determines the rotor current, I_2, and hence the output torque. In this mode of operation, the peak torque which can be developed, is constant throughout the speed range, and the motor is said to possess a constant-torque characteristic. If the stator leakage impedance voltage drop is small, the terminal voltage, V_1, is increased linearly with frequency in order to maintain constant air-gap flux. However, the stator voltage can only be increased up to the maximum output voltage of the converter. This is determined by the voltage ratings of the thyristors in the converter circuit. The speed range can be extended above this base speed if the converter voltage is held constant at its limiting value as the frequency is increased. This causes the air-gap flux to decrease with frequency, and the motor torque-speed characteristics are modified as in Fig. 6.10. By suitable control of the rotor frequency, f_2, in a controlled-slip

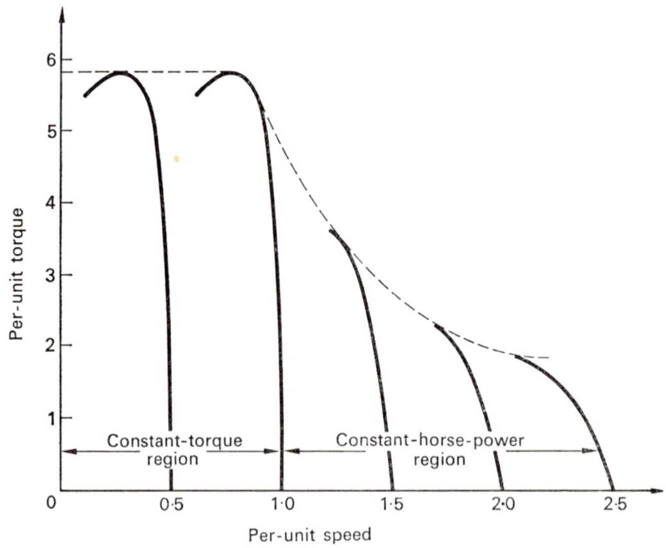

FIG. 6.10. Typical torque-speed characteristics for variable-frequency operation of the induction motor.

drive, the output torque can be made to vary inversely with speed, thus providing a limited constant-horsepower range of operation. This corresponds to the field-weakening operation of a d.c. shunt motor.

In general, the induction motor torque at small slip is given by equation (6.18):

$$T = K''\Phi^2 f_2$$

and hence

$$T \simeq K''\left[\frac{V_1}{f_1}\right]^2 f_2.$$

For a constant stator voltage, V_1, and rotor frequency, f_2, the torque decreases as the square of the supply frequency, f_1. However, if f_2 is increased linearly with f_1, the torque is inversely proportional to the supply frequency. This means that the output torque varies approximately as the inverse of the speed, and hence a constant-horsepower characteristic is obtained. The upper speed limit is determined by the maximum working value of rotor frequency. In practice, the rotor frequency is limited to about half the breakdown frequency at the upper speed limit. Operation nearer the breakdown point produces increased currents and increased copper loss without a significant increase in torque output. This limits the constant-horsepower range to about 2·5 times the base speed.(149)

A constant-horsepower characteristic may also be obtained by operating with a constant slip frequency and controlling the stator voltage and frequency together, so that the required torque is developed.

Again,

$$T = K''\Phi^2 f_2$$

and, for a fixed value of f_2,

$$T \propto \Phi^2 \propto (V_1/f_1)^2.$$

If V_1^2 is proportional to f_1, the torque is inversely proportional to f_1, and a constant-horsepower output is obtained. Thus, the voltage variation is proportional to the square root of the frequency range. This type of drive behaves like a d.c. series motor with speed control provided by variation of the motor voltage.(150) For traction applications which require constant-horsepower operation over a wide speed range, this method can be used at low speeds and the constant-voltage method at higher speeds.

6.7.2. *Advantages of the Controlled-slip Drive*

The controlled-slip induction motor provides a highly efficient drive with precise contro of torque over a wide speed range down to standstill. A large torque can be obtained at high power factor and high efficiency by operating at rotor slip frequencies below the breakdown value. Stability of operation is ensured by incorporating a slip limit. Since the induction motor is not run-up from standstill by the step application of mains frequency, there is no necessity for a high-resistance rotor winding to limit starting current and increase starting torque. By choosing a low-slip motor with a breakdown slip of 4 per cent or less, the rotor losses are reduced to a minimum under normal operating conditions. This offsets the additional losses due to harmonic currents in the rotor winding. Regenerative braking can be incorporated, as already explained, and consequently closed-loop control of absolute slip frequency provides a variable-speed drive with all the flexibility of a d.c. motor in a Ward–Leonard or double-converter system. In fact, the controlled-slip drive provides high efficiency operation over a wider speed range than is possible with thyristor-controlled d.c. drives. The cage-rotor machine also has the advantages of reliable maintenance-free operation for prolonged periods due to the elimination of all brush contacts. The induction motor also avoids the limitations on rotational speed imposed by the restricted peripheral speed of the d.c. commutator. In addition, the cage-rotor motor has a smaller moment of inertia for a given horsepower than any other machine, and a better dynamic performance is possible.

PLATE 1
A fractional horsepower a.c. motor drive. The inverter has a rated output of 4 kVA and the operating frequency range is 20 to 120 Hz.
(Courtesy of General Electric Co., U.S.A.)

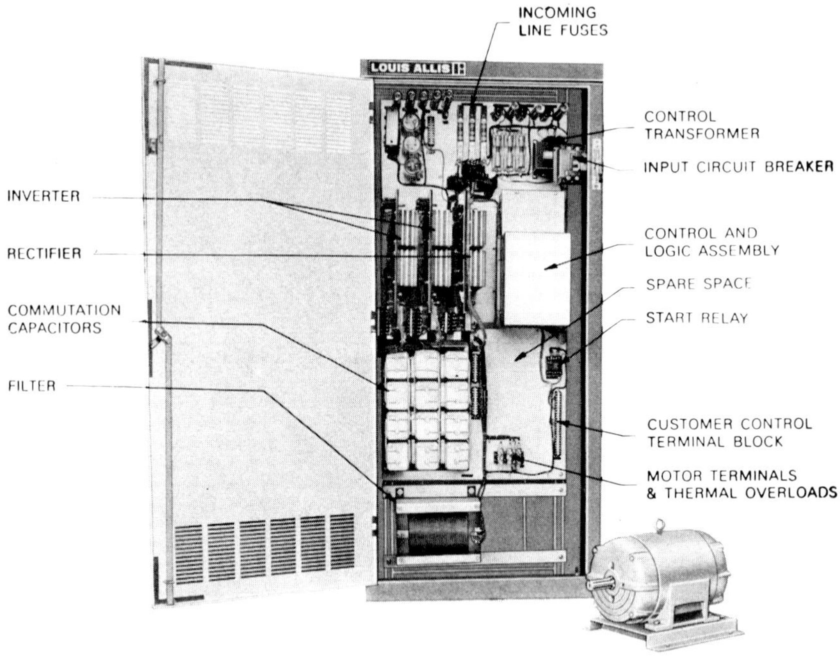

PLATE 2
A 20 horsepower a.c. motor drive showing the cabinet layout of the variable-frequency controlled-current d.c. link converter.
(Courtesy of Louis Allis Co., Milwaukee, U.S.A.)

PLATE 3
A 60 kVA variable-frequency d.c. link converter for a.c. motor speed control. The output frequency range is 2 to 200 Hz.
(Courtesy of A.E.G., Germany)

CHAPTER 7

The Static Variable-frequency A.C. Drive

7.1. INTRODUCTION

The basic circuitry and modes of operation of the solid-state variable-frequency a.c. drive have been introduced in preceding chapters. The cycloconverter is attractive for low-speed drives, but the d.c. link converter offers greater flexibility and has been more widely used in industry. The selection or specification of a d.c. link converter and the selection of a suitable a.c. motor are now considered. The precautions necessary to ensure satisfactory and reliable operation of thyristor circuits are also briefly treated. The problem of harmonic distortion is introduced, and the all-important economic considerations are reviewed.

7.2. STATIC INVERTER RATING

A thyristor rectifier or inverter has limited ability to withstand overloads, and care must be exercised when specifying or selecting a converter for an a.c. motor drive. The thyristor itself has a precise maximum junction temperature which cannot be exceeded without damaging the device. This necessitates an efficient cooling system to remove the heat generated by internal losses, principally the power loss in the forward-conducting state. The thyristor is normally mounted on a large metal fin or heat sink which conducts heat away from the temperature-sensitive junction. The heat is then dissipated to the surrounding air by radiation and convection. In high-power installations, forced-air cooling or water cooling may be used.

The forced-commutated inverter has three current ratings, each of which must be strictly observed. The continuous current rating is determined by the ambient temperature, the thermal resistance of the heat sink, and the maximum allowable rise in thyristor junction temperature. The small thermal capacity of a thyristor causes it to heat up rapidly on overcurrents, and the overload capability of a static inverter is much less than that of an electric motor, which has a thermal time constant measured in minutes. Consequently, in a motor, large overcurrents may be carried for short intervals of time without causing overheating, but in a thyristor, destructive overheating occurs in a fraction of a second, and the overcurrent rating based on the transient thermal characteristics must be carefully respected. A third current rating is also specified for the inverter. This is the instantaneous peak current delivered by the inverter and is limited by the commutating capability of the system. If the current at the instant of commutation exceeds this limit, a commutation failure occurs, which may blow fuses or trip out the inverter. The maximum current to be commutated is decided at the design stage, and the commutating circuit is constructed on this basis. In subsequent service, the built-in commutating capability of the inverter cannot be exceeded without causing a shut-down.

The static inverter has, therefore, got three kVA ratings—the continuous, momentary and commutating ratings. The continuous kVA rating is the maximum output which the inverter can deliver indefinitely, at maximum ambient temperature, while remaining within the frequency and voltage specifications. The momentary kVA rating is the maximum inverter output for a specified time, following operation at rated continuous current and maximum ambient temperature. The commutating kVA rating is the peak instantaneous output which the inverter can deliver without exceeding the commutating capability of the system. A typical inverter momentary rating is 150 per cent of the continuous rating for one minute, and 200 per cent of the continuous rating for 0·5 second. Unless the commutating rating is specified independently, it is assumed to be the same as the momentary rating.

The momentary rating of a static inverter is considerably less than that normally associated with rotating power supplies, and this is a serious disadvantage in a.c. motor applications. When started at rated voltage and frequency, a typical squirrel-cage motor draws six times full-load current, while a reluctance motor may draw up to eight times rated current. If the static inverter is to supply these large transient currents, it must have a continuous rating of this order, even though it only delivers the currents for the run-up period. The need to over-size the inverter increases the cost of a static a.c. drive and may render it uneconomic for some applications. A current limit is often incorporated in the system in order to keep the motor current within the inverter rating. This also provides overload protection, but it limits the accelerating torque when starting and when a rapid speed increase is demanded. Improved starting performance is obtained with a reasonable inverter rating if the voltage and frequency can be reduced at starting, and then increased gradually in order to speed-up the motor. However, in multi-motor drives it is not always feasible to start the motors simultaneously from standstill. If the motors are run-up individually, the inverter must have a momentary current rating equal to the standstill current of one motor, plus the full-load current of the remainder.

It must also be borne in mind that the continuous current rating of the inverter varies with frequency. At very low frequencies, each thyristor carries the full input current for an appreciable interval, and at zero frequency the direct current rating of the inverter equals the continuous current rating of one thyristor. As the operating frequency increases, the conduction periods become shorter and the current rating increases. Even at frequencies of 1 or 2 Hz, the inverter current rating is about 40 or 50 per cent greater than the d.c. rating. However, circulating currents are also present in some inverter circuits due to the commutation method employed. The mean circulating current increases with output frequency, and consequently the continuous rating is reduced at high frequencies.

Most inverter circuits generate a non-sinusoidal output voltage, and the harmonic distortion of the inverter current must be taken into consideration. The presence of harmonics increases both the r.m.s. and peak values of the inverter current. The increase in r.m.s. value causes additional heating and may necessitate somewhat larger cables and heat sinks. The increase in peak inverter current imposes a more severe commutating duty on the inverter.

As already explained in Chapter 5, the harmonic currents are largely determined by the leakage reactance of the a.c. motor and are practically independent of the loading conditions and speed. The static inverter is frequently designed to generate a six-step or twelve-step voltage waveform, and Fig. 5.3 of Chapter 5 plots the total r.m.s. inverter current at full-load as a function of the per-unit reactance of the motor. In Fig. 5.6 the ratio of the peak inverter current to the peak fundamental current at full-load is shown. The inverter rating is usually expressed in r.m.s. units based on a sine wave, and the curves in Fig. 5.6 also express the ratio of the peak r.m.s. current to the fundamental r.m.s. current.

7.2.1. *Static Inverter Specification*[151-2]

The induction motor and reluctance motor operate at lagging power factor, and the reactive power requirements must be taken into consideration when specifying the inverter rating. The rectifier which supplies d.c. power to the inverter is only rated for the real power, but the inverter itself must be rated for the apparent power. The inverter frequency range is determined by the desired speed range, the number of poles in the a.c. motor, and the gear ratio, if any. With small a.c. motors, the minimum frequency is often about 10 Hz, in order to prevent overheating due to reduced ventilation and also to avoid the speed pulsations and motor instability which may occur at lower frequencies. The maximum inverter frequency with standard a.c. motors is often in the region of 150 or 200 Hz, since this avoids excessive core loss in the motor and excessive commutating losses in the inverter.

The a.c. motor usually delivers a constant-torque output, and the stator voltage increases linearly with frequency, while the stator current is roughly constant over the frequency range. Thus, the motor kVA is greatest at the highest speed of operation, and the continuous kVA rating of the inverter is calculated at the maximum output frequency. In order to determine the inverter rating required in a given application, the number of motors must be specified, and their full-load and locked-rotor characteristics on a sine-wave supply must be known. The continuous kVA rating of the inverter is determined by multiplying the kVA per motor at maximum frequency by the number of motors employed in the drive, and also by the appropriate harmonic factor obtained from Fig. 5.3. This factor takes account of the increase in r.m.s. inverter current due to the presence of harmonics. The continuous kVA rating should also be checked at the minimum frequency, if the inverter has a voltage boost for low-frequency operation.

The following example illustrates the specification procedure. Specify the kVA ratings of a six-step inverter to supply twenty-four three-phase motors, each having a full-load rating at the maximum frequency of 2 A at 200 V and 0·6 power factor. The corresponding locked-rotor current of the motor at rated voltage is 16 A at 0·5 power factor. Four motors are required to start simultaneously.

The per-unit reactance of the motor is calculated by substituting in equation (5.17).

Thus,

$$X_{\text{pu}} = \frac{I_1}{I_s} \sin \phi_s = \frac{2}{16} \sin (\text{arc} \cos 0{\cdot}5)$$

$$= 0{\cdot}108.$$

Figure 5.3 shows that the total r.m.s. current at full-load is 1·09 times the fundamental r.m.s. value. The continuous kVA rating of the inverter is, therefore,

$$kVA_{cont} = 24\,(\sqrt{3} \times 200 \times 2 \times 10^{-3}) \times 1{\cdot}09 = 18{\cdot}1 \text{ kVA}.$$

The momentary rating of the inverter must be sufficient to supply the standstill current of four motors together with the full-load current of the remainder.

Hence,

$$\begin{aligned} kVA_{mom} &= 4\,(\sqrt{3} \times 200 \times 16 \times 10^{-3}) + 20\,(\sqrt{3} \times 200 \times 2 \times 10^{-3}) \times 1{\cdot}09 \\ &= 37{\cdot}3 \text{ kVA}. \end{aligned}$$

The time for which the momentary rating must be maintained depends on the starting duty. In this calculation, the locked-rotor current is not multiplied by the harmonic factor, since the fundamental motor current at standstill is eight times rated current, but the harmonic currents are practically the same at full-load and at standstill. Consequently, the locked-rotor current is mainly a fundamental current.

The peak or commutating rating must be sufficient to deliver the maximum instantaneous inverter current. This also occurs when starting four motors with all other motors already running, but the commutating rating is greater than the momentary rating because it takes account of the peak inverter current due to harmonic distortion. From Fig. 5.6 the peak r.m.s. current is 1·78 times the fundamental r.m.s. current. Thus,

$$\begin{aligned} kVA_{comm} &= 4\,(\sqrt{3} \times 200 \times 16 \times 10^{-3}) + 20\,(\sqrt{3} \times 200 \times 2 \times 10^{-3}) \times 1{\cdot}78 \\ &= 46{\cdot}8 \text{ kVA}. \end{aligned}$$

Again, in this calculation, the locked-rotor current is not multiplied by the harmonic factor, since the current waveform is nearly a true sine wave. Each of the above ratings is normally multiplied by a factor of 1·1 or 1·2. This allows a 10 or 20 per cent safety margin for manufacturing tolerances and unbalance between phases.

The continuous rating of the rectifier is determined from a knowledge of the continuous kVA rating of the inverter and its output power factor and efficiency. The product of inverter kVA and power factor gives the kW output, which is divided by the efficiency in order to determine the input power to the inverter. The commutating losses are the principal losses in the inverter, and an efficiency of 80 per cent can be assumed as a conservative estimate. When several inverters are operating from a common rectified supply, the rectifier rating is less than the sum of the peak loads of all inverters, since each inverter will not operate at maximum frequency and full-load at the same time. With five or more inverters, a reduction factor of 0·8 can be assumed.

7.3. MOTOR SPECIFICATION[153–4]

Solid-state a.c. motor drives usually employ squirrel-cage induction motors or reluctance motors. Standard motors are frequently used, but for optimum performance, special machine designs are necessary. Co-ordination of the motor and the power supply is the important factor in providing the required operating characteristics and ensuring that the system remains stable for all operating conditions. Parasitic torque and speed pulsations may also be minimized by a suitable choice of system parameters. To optimize the design, a computer simulation of the complete system is necessary at the design stage. This permits a choice of suitable circuit values and motor parameters. A motor whose design is optimized in this manner for variable-frequency operation may differ considerably in its design parameters from a standard motor intended for fixed-voltage, fixed-frequency operation.

The speed of the squirrel-cage induction motor normally lies within a few per cent of synchronous speed, and this is adequate in many applications where very precise speed control is not essential. If the supply voltage and frequency can be reduced for starting, the normal direct-on-line conditions are not applicable, and the problem of excessive starting current does not arise. The induction motor is, therefore, designed with a smaller-than-usual rotor resistance and, in this manner, a speed regulation from no-load to full-load of 1 or 2 per cent can be obtained. Improved performance may be achieved by employing a slip-compensation technique in which the stator frequency is increased as load is applied to the motor. Alternatively, conventional closed-loop methods of speed control may be used. In multiple-motor conveyor drives, high-slip induction motors with 5 to 8 per cent slip at full-load are employed in order to improve the load sharing between individual motors.

The reluctance motor provides very precise speed control in a simple open-loop system. Digital control techniques permit speed-holding accuracies of 0·001 per cent, or better, at any set speed. As explained in Section 1.5.2, the important machine parameters are the pull-in torque, the pull-out torque, the power factor and the starting current. The pull-in torque is dependent on the load friction and load inertia, but the pull-in performance is improved if a low-resistance rotor winding is fitted. This allows the motor to run-up very close to synchronous speed as an induction motor, and a high inertia load can then be accelerated into synchronism. However, the provision of a low-resistance cage winding also increases the standstill current drawn by the motor, and this can be particularly objectionable when the supply is obtained from a thyristor converter. Large pull-in torque with high inertia also reduces the power factor and efficiency of the motor, giving a poor synchronous performance. For normal operation on a fixed-frequency supply, the motor design is a compromise between these factors. If, however, the motor is supplied with a variable frequency which is reduced for starting and then increased gradually, the motor can be designed for optimum synchronous performance without regard to the pull-in requirements.

7.3.1. *Harmonic Losses and Motor De-rating*

For variable-frequency operation, it is not feasible to filter the output of the static converter and, consequently, harmonic effects may be significant. As shown in Chapter 5, the harmonic currents have negligible effect on the average torque developed by the motor, but pulsating torques are also developed, and these produce a non-uniform stepping motion of the rotor at low speeds. Motor instability may occur, and the techniques for eliminating it have been described in Chapter 5. The harmonics in the motor current may also cause increased acoustic noise due to magnetostriction, but in a well-designed machine the additional noise is not significant. Harmonic effects are accentuated when voltage control is obtained by simple pulse-width modulation or phase-shift control.

Excessive pull-in torque should not be specified in a reluctance motor for a static variable-frequency a.c. drive, since this will necessitate a substantial cage winding with a low leakage reactance. As a result, the motor will draw appreciable harmonic currents, and the harmonic copper loss may be excessive. As already stated, an induction motor in a variable-frequency a.c. drive should have a small rotor resistance in order to reduce the speed regulation and improve the motor efficiency on load. The harmonic copper loss in the rotor is one of the main causes of reduced efficiency on non-sinusoidal supplies, and a low-resistance rotor will significantly reduce this contribution to the harmonic losses. The use of an unskewed rotor construction is also desirable in order to eliminate the skew-leakage harmonic loss. Machine design can be further optimized for minimum losses using a digital computer, or the supply conditions for a standard motor can be optimized to yield the required torque and speed with minimum motor losses.[144,155] However, at frequencies in excess of 150 or 200 Hz, the core loss in a standard machine is usually excessive, and this sets a limit to the maximum frequency of operation.

The additional harmonic losses must be taken into consideration when the a.c. motor operates on a non-sinusoidal supply. As shown in Section 5.4, the reduction in motor efficiency is usually not excessive, since the increase in losses is less than 20 per cent. However, in order to avoid overheating it is necessary to reduce the continuous power and torque ratings of the motor. The required de-rating depends on the degree of harmonic distortion. With a six-step voltage waveform, having a fundamental component equal to the rated sine-wave voltage, most a.c. motors draw a total r.m.s. current which is not more than 10 per cent greater than the rated sine-wave current. This causes an increase of up to 20 per cent in stator copper loss and the motor temperature is up to 5°C higher than on a sine-wave supply. As a result, the motor rating must be reduced by up to 10 per cent of its sine-wave rating. With a high-reactance motor the de-rating is less, and with a twelve-step voltage supply the reduction in motor rating is usually negligible.

Voltage control methods such as phase-shift control and simple pulse-width modulation produce excessive harmonics at low output voltages, and this necessitates considerable de-rating of the a.c. motor unless steps are taken to reduce the current and/or voltage harmonics. Harmonic currents are attenuated by designing the motor with a large leakage inductance,[99] or by inserting external series inductors in the stator leads of the motor.[82] The inductive reactance presented to the harmonic currents is greater than that offered to the fundamental current, and hence the current distortion is reduced and the operating efficiency is increased. On the other hand, the fundamental breakdown torque is reduced. Voltage distortion at low output voltages may be reduced by means of a step-down transformer in the inverter output, as explained in Section 3.6.3. More complex systems of multiple-pulse-width modulation, in which the pulse width is sinusoidally modulated during each half-cycle, have a reduced harmonic content, and auxiliary transformers and series inductors are not required. Such systems provide a near-sinusoidal output voltage waveform, but the control circuitry is more complex and costly.

The pulse-width modulated inverter is also readily modified to deliver a constant-current supply to an a.c. motor. In this mode of operation, the a.c. motor is supplied with a sinusoidal or six-step current waveform of controlled amplitude. The inverter thyristors are rapidly switched on and off so that the output current oscillates about the prescribed value.

As explained in Section 6.4, an inverter operating under controlled-current conditions has built-in overcurrent protection and the commutating circuitry is designed to commutate the maximum rated current. This gives a higher operating efficiency and permits better utilization of the inverter during normal operation, but complex control circuitry is again necessary.

7.3.2. *Motor Characteristics at Variable Frequency*

The torque-speed characteristic of the a.c. motor is determined by the output voltage-frequency characteristic of the static converter, as explained in Chapter 6. A constant volts/Hz output gives an approximately-constant air-gap flux density and maintains the shaft torque nearly constant over the speed range. Thus, a 440-V, 50-Hz motor requires 220 V at 25 Hz and 880 V at 100 Hz. Resistance effects become significant at low frequencies and the air-gap flux is reduced. In order to maintain the output torque, the stator volts/Hz must be increased progressively when the frequency is reduced below about 20 Hz. The required volts/Hz boost depends on the machine size and design, but a 10 per cent increase is typical at 15 Hz and a 20 per cent increase at 10 Hz.

With a constant-torque output, the horsepower rating of the machine is proportional to speed, and large horsepower is available at high speed. Thus, a motor which is rated for 5 h.p. at 1000 rev/min will deliver 20 h.p. at 4000 rev/min, provided the motor is able to dissipate the extra power loss. Improved ventilation at high speeds may, in fact, permit increased current densities. The horsepower rating of the machine then increases more rapidly than the speed, and compact high-power motors are feasible when high-frequency operation is possible. The current rating at a particular speed depends on the cooling method and type of enclosure used. Reduced ventilation at low speeds may necessitate some reduction in the motor rating.

For a constant-torque output, the motor flux should be maintained at high speeds by increasing the stator voltage linearly with frequency. The applied voltage is limited by the motor insulation and by the peak voltage rating of thyristors in the converter circuit. With an induction motor, a constant-horsepower mode of operation is sometimes provided at speeds above the constant-torque range, by operating the motor with a constant terminal voltage (Section 6.7.1).

As already stated, the reluctance and induction motors are the most common types used with inverter supplies. If a high starting torque is required, an induction motor is preferable to a reluctance motor. Permanent-magnet synchronous motors are also used, but they are expensive. Conventional d.c.-excited synchronous motors may be employed in high-power applications where the slip-rings are not objectionable. These machines have the advantage of power factor control by adjustment of the d.c. field excitation. This facility can be used to reduce the inverter rating and also to ease the commutating requirements.

When an adjustable-frequency method is used, the machine design can be optimized by an appropriate choice of operating frequency. Conventional low-speed a.c. or d.c. motors have a large number of stator poles and, consequently, they are bulky and expensive. They also have a large inertia which results in a poor dynamic performance. In an adjustable-frequency a.c. system, a low operating frequency may be chosen in order to obtain a compact low-speed motor with a small rotor diameter and large axial length. This construction gives a reduced inertia and improved dynamic performance. For high-speed operation, a higher operating frequency is selected and a compact high-power motor is obtained.

7.4. THYRISTOR CIRCUITRY

In the various thyristor circuits illustrated in preceding chapters, only the essential components were included to avoid confusion. In a practical installation, auxiliary components are necessary to protect thyristors from the effects of overvoltages and overcurrents, and also to limit the rate of rise of forward blocking voltage (dv/dt) and the rate of rise of anode current (di/dt). Voltage-sharing components are required with a series chain of thyristors in order to ensure that transient and steady-state voltages are properly distributed. Similarly, a group of parallel thyristors requires current-sharing components to distribute the current evenly among the group.

7.4.1. *Overvoltage Protection*

Like most semiconductor devices, the thyristor is very sensitive to excessive voltage, and transient overvoltages of very brief duration may cause destructive breakdown. Every thyristor has a peak reverse voltage rating which cannot be significantly exceeded without damaging the device. (Some thyristors have a controlled avalanche characteristic in the reverse direction which prevents thyristor destruction on brief reverse voltage transients.) If the forward breakover voltage, $V_{(BO)}$ is exceeded, the thyristor switches into the forward-conducting region. This may result in thyristor damage if the peak forward voltage rating is less than the breakover voltage. In Appendix B it is also explained that thyristor firing may occur when the forward blocking voltage is very rapidly increased, even though there is no gate signal and the forward breakover voltage is not exceeded. This is known as dv/dt turn-on. Random triggering by forward breakover or excessive dv/dt causes the circuit to malfunction, and large fault currents may flow as a result. Consequently, transient forward or inverse overvoltages may cause thyristor damage and circuit interruption. For reliable circuit operation it is important to limit transient overvoltages by adopting adequate suppression techniques.

Origin of voltage transients. Voltage transients may originate within the thyristor circuit or they may be superimposed on the normal a.c. mains voltage due to system switching operations and occasional lightning surges. If a thyristor converter is supplied by a mains transformer, a voltage transient is generated whenever the primary is suddenly energized or de-energized. When the primary is energized, a ringing oscillation occurs on the secondary side due to the sudden application of voltage to the oscillatory circuit formed by the secondary leakage reactance and winding capacitance. If the primary is energized at a peak of the supply voltage, an overvoltage of twice the supply voltage can be generated. The voltage transient following sudden interruption of the primary magnetizing current is usually the most severe transient occurring in practice, as the main transformer flux is forced to decay rapidly to zero. The interruption of current flow in an inductive circuit is always liable to produce dangerous overvoltages, unless alternative low-resistance discharge paths are provided. The commutation or turn-off of a thyristor may generate overvoltages which are capable of damaging the thyristor itself, or other semiconductor devices in the circuit. These are called hole-storage or commutation transients and are caused by the rapid decay of the hole-storage or reverse recovery current (Section 2.2). Current changes of 10^6 or 10^7 A/sec occur in practice, and may generate a significant overvoltage if flowing through a series inductance such as a stray circuit inductance or transformer leakage inductance. The rapid current change also generates considerable radio-frequency interference (Section 7.6).

Suppression of voltage transients. In a practical installation it would be uneconomical and impractical to use thyristors which are capable of withstanding all transient overvoltages, as these may rise to more than ten times the normal working voltage. In practice, thyristors are chosen with a peak voltage rating of about 2·5 times the normal peak working voltage. With properly designed transient suppression circuits, this leaves sufficient margin for supply voltage fluctuations and residual transients after suppression. The voltage transients are usually suppressed by means of capacitive networks or selenium surge suppressors.

A shunt capacitor connected across the thyristor is the simplest form of transient suppression. The slow charging rate of the capacitor limits the magnitude of fast voltage transients and reduces the rate of forward voltage rise on the thyristor. A small series resistor is included with each capacitor to damp ringing oscillations between the capacitance and stray circuit inductance. This is called a snubber circuit. The damping resistor also limits the initial inrush current when the thyristor is triggered (Section 7.4.3). Additional resistance-capacitance networks are often introduced between the secondary lines of the input transformer, and empirical formulae are available giving conservative component values.

The selenium surge suppressor is now marketed under various trade names. It consists of a selenium rectifier with a well-defined reverse breakdown region in which a small increase in voltage produces a considerable increase in current. The characteristic is, therefore, similar to that of a zener diode. At normal voltages the selenium cell draws a small reverse leakage current. Large transient voltages exceeding the breakdown value are clipped by an excursion into the breakdown region of the characteristic. The surge energy is dissipated in the rectifier, but the operating point returns to its original position after the surge, provided excessive energy has not been dissipated.

7.4.2. *Overcurrent Protection*

The small thermal capacity of a silicon diode or thyristor causes the junction temperature to respond very rapidly to changes in current. If the semiconductor circuit is fed by a weak supply system, the fault current is limited by the source impedance, and semiconductor damage may not occur for several cycles. In such systems, the fault current can be interrupted by conventional fuses or circuit-breakers. Proper co-ordination is essential to guarantee that the fault current is interrupted before semiconductor damage takes place, and to ensure that only faulty branches of the circuit are isolated. Overcurrent protection may also be provided by removing the gating signals when an overcurrent is detected. However, these protection methods are inadequate in thyristor circuits which are supplied by an electrically stiff system. In such applications, the junction temperature rises to an excessive value within a few milliseconds and the semiconductor is destroyed. Special high-speed current-limiting fuses are required for the protection of solid-state devices in these circuits.

Current-limiting fuses.[156] These fuses have similar thermal properties to the silicon diode or thyristor, and this simplifies the co-ordination of fuse and semiconductor. The fusible link consists of one or more fine silver wires of small thermal capacity mounted between heavy metal end-pieces. This is thermally analogous to a solid-state device mounted on a cooling metal heat sink. The light silver elements have a very short fusing time, giving the fuse a current-limiting action by forcing the available fault current to zero before it rises to its peak value. In Fig. 7.1, the available fault current is indicated, assuming the circuit

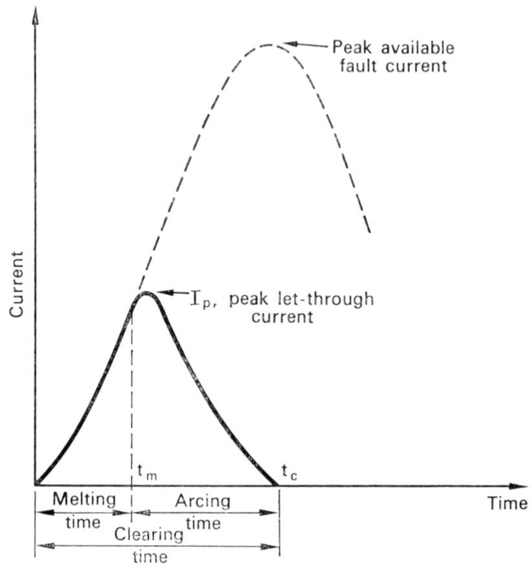

FIG. 7.1. The limitation of fault current in an a.c. circuit by means of a current-limiting fuse.

current prior to the fault is negligible. The fuse impedance is also negligible, and when a fault occurs at $t = 0$ the current increases in the usual manner until a time, t_m, when the fuse melts. An arc is then struck, but the current may continue to rise for a brief interval depending on the circuit parameters and the fuse design. The maximum current, I_p, is called the peak let-through current and is considerably less than the peak available current. This shows the current-limiting action of the fuse.

The fault current decreases from its peak value as the arc resistance increases. The energy stored in the circuit inductance during the build-up of fault current is dissipated in the arc loss until eventually, at time t_c, all the stored energy has been dissipated and the arc is extinguished. The total clearing time, t_c, is the sum of the melting time, t_m, and the arcing time, $(t_c - t_m)$, as in Fig. 7.1.

Fuse melting occurs when the energy dissipated in the fusible element raises its temperature to the melting value. At any instant, let i denote the fault current and R the resistance of the fusible link. The instantaneous power input to the fuse is i^2R and the total energy input during the melting time is $\int_0^{t_m} i^2 R\, dt$. The value of $\int_0^{t_m} i^2\, dt$ is an important parameter in fuse behaviour and is usually termed the melting I^2t of the fuse. In many fuse designs it is approximately constant for millisecond melting times. The total clearing I^2t is given by $\int_0^{t_c} i^2\, dt$. This is also described as the let-through I^2t of the fuse, and is a measure of the total thermal energy which the semiconductor devices must withstand before the fault current is interrupted.

Fuse selection. As stated above, the current-limiting fuse and semiconductor device have the same thermal properties. Consequently, for times in the millisecond range, the semiconductor is also described in terms of its fusing I^2t value. A fuse is usually placed in series with each semiconductor, and obviously must be rated to carry the full-load current indefinitely. Overcurrent protection is provided by selecting a fuse whose let-through I^2t rating is less than the I^2t rating of the semiconductor which it protects. This simple co-ordination procedure eliminates the necessity for estimating fault currents for millisecond intervals. Additional fuses may be inserted in the supply leads. In practice, the fuse let-through current and I^2t is a function of the available fault current, the circuit voltage, the circuit X/R ratio and also the point in the voltage cycle at which arcing commences. However, a conservative value of clearing I^2t is usually specified in order to co-ordinate the fuse and semiconductor.

The voltage across the fuse during the arcing period is known as the arcing or recovery voltage, and equals the sum of the source voltage and the induced e.m.f. in the circuit inductance. If the fault current is interrupted very rapidly, the arcing voltage may attain excessive values due to the inductive e.m.f., $L\, di/dt$. By careful fuse design, however, the arcing voltage may be limited to less than twice the peak supply voltage, and this type of fuse is suitable for semiconductor applications.[157] The voltage rating of the fuse should not be greatly in excess of the circuit voltage, or this will result in excessively abrupt current interruption and destructive overvoltages. On the other hand, if the voltage rating of the fuse is inadequate, the arcing time and let-through I^2t are both increased.

In general, semiconductor circuits may require conventional fuses or circuit breakers for less severe faults, in addition to the current-limiting fuse which only operates positively on severe overloads. A co-ordination chart can be drawn to ensure that adequate protection is provided at all overcurrent levels.

Proper fuse co-ordination is also important in a multi-motor drive so that when a fault occurs in one motor, inverter operation is uninterrupted, and only the faulty machine is disconnected by its protective fuses. This co-ordination is achieved if the peak let-through current of the motor fuse does not exceed the commutating current rating of the inverter. Hence, in this application, the motor fuse must be chosen with regard to peak let-through current rather than let-through I^2t.[158]

7.4.3. *Limitation of Rate of Rise of Anode Current* (di/dt)

When a thyristor is triggered into conduction by a gate signal, there is a finite turn-on time during which the anode voltage decreases from the blocking value to the forward-conducting value. Simultaneously, the anode current increases, and the product of voltage and current represents the instantaneous power loss in the thyristor during turn-on. If the anode current increases rapidly, the power loss at turn-on becomes significant, and overheating may occur at high switching frequencies. The problem is aggravated by the fact that the anode current is initially confined to a small area near the gate electrode. The conducting area then expands over the rest of the junction, but if the anode current increases rapidly, the power loss is initially dissipated in a localized hot-spot near the gate electrode. Thus, excessive rate of rise of forward current (di/dt) will damage the thyristor, and a critical di/dt value is specified.

Excessive di/dt may occur when a thyristor which blocks a high forward voltage is suddenly triggered into a resistive or capacitive load, but the slow build-up of current in an inductive load prevents excessive inrush current. When thyristors are fitted with shunt capacitors for transient voltage suppression or voltage sharing, a small series resistor is included to limit the inrush current when the thyristor is triggered. Small series inductors are sometimes included in high-power inverter circuits to limit the initial di/dt. Saturable reactors are also used to delay the build-up of current for a few microseconds. The reactor then saturates and it no longer affects the circuit operation.

7.4.4. *Series and Parallel Operation of Thyristors*

When circuit applications require voltage or current ratings in excess of those provided by a single thyristor, it becomes necessary to connect several thyristors in series or in parallel. Certain precautions must be taken to ensure satisfactory operation under these conditions.

Series operation. For high-voltage applications, several thyristors are connected in series, and auxiliary components are necessary to ensure that the applied voltage is uniformly distributed over the thyristor chain (Fig. 7.2). Variations in the reverse and forward-blocking characteristics can produce an uneven steady-state distribution. Thus, if one thyristor in the series chain has a very high leakage resistance, it supports most of the applied voltage, and its peak voltage rating may be exceeded. A more uniform voltage distribution can be obtained by connecting a resistor across each thyristor. The value of this shunt resistance is calculated by assuming a "worst-case" condition in which one thyristor in the chain has negligible leakage and all others have maximum leakage current. The resistance is usually chosen to equalize the forward blocking distribution, as the currents are generally somewhat greater than the reverse leakage currents.

The resistive potential divider is satisfactory for steady-state conditions, but variations in thyristor recovery time can cause an uneven transient distribution. If one thyristor in the chain has a short recovery time, it has to support the full applied voltage for a brief interval while the remaining series thyristors recover their reverse-blocking capability. A capacitive potential divider improves the dynamic voltage sharing, since the slow charging of the shunt capacitor limits the build-up of voltage on the first thyristor to recover. The capacitive divider also forces voltage sharing during sudden changes in applied voltage, since the external capacitance swamps variations in junction capacitance. A small series resistor is included with each capacitor in order to damp out oscillations with the distributed circuit inductance.

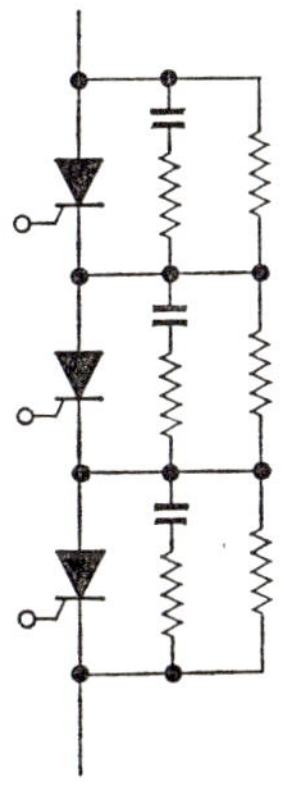

FIG. 7.2. Series connection of thyristors, with auxiliary components to equalize the steady-state and transient voltage distributions.

Parallel operation. Thyristors are operated in parallel when the circuit current exceeds the rating of an individual thyristor. The forward characteristic of the thyristor in the conducting region has a very low slope resistance, and a slight change in voltage produces a large variation in current. When two thyristors are connected in parallel, the voltage drop must be the same for both, and widely different currents may flow as a result. Selected thyristors having matched characteristics should be used for parallel operation. The current rating of the combined unit is then only 10 per cent, or less, below the sum of the individual current ratings.

Forced current-sharing is necessary with unmatched thyristors. Series resistors can be used to increase the effective slope resistance and thereby reduce the current difference. This is obviously inefficient but the method is simple and economical for low-power applications. Current-balancing reactors are a more efficient method of forced current-sharing, but for low-frequency operation the reactors are bulky and expensive.[159]

7.5. CONTINUITY OF OPERATION

Continuity of supply is essential in many critical process industries in order to avoid production losses and damage to plant. In such applications, it is vital that the drive should ride-through a.c. supply disturbances and interruptions. For non-critical loads, an orderly shut-down of plant is satisfactory, and the system is then ready for an immediate restart when power is restored.

The static inverter has a large capacitor across the d.c. supply to filter the input voltage and to provide the high peak currents which some circuits require at commutation. If the d.c.

link voltage falls excessively, the inverter fails to commutate properly and must be shut-down before fuses are blown. When the d.c. link supply is interrupted as a result of an a.c. power failure, the smoothing capacitor rapidly discharges, and the inverter output is maintained for only a few cycles of the network frequency. Protective circuitry disconnects the inverter when the commutating voltage is insufficient. Auxiliary capacitor banks can be connected across the d.c. supply to extend the inverter shut-down time to twenty cycles or so. This provides continuity of supply for short-duration voltage dips. Motor-generator sets with flywheels are also used to provide short-term protection. When the a.c. supply fails, the flywheel continues to rotate for a few seconds and the generator maintains the d.c. link supply. For long-term stand-by capacity, however, it is essential to use an auxiliary battery supply. This may be connected as shown in Fig. 7.3. A simple diode rectifier bridge is used for reliability and, in the event of an a.c. power failure, the inverter d.c. supply is obtained from the stand-by battery. In this manner, the inverter operation is maintained for 30 minutes or more, depending on the battery capacity. This provides continuity of supply until a.c. power is restored.

Static inverters are sometimes operated in parallel in order to provide large amounts of power or to give increased reliability in the event of a failure of one inverter. Before the inverters are paralleled, the usual synchronizing requirements of equal voltages, equal frequencies and zero phase difference must be satisfied. The master oscillators in the inverter control circuits are also locked in synchronism before paralleling the inverters. This is necessary to prevent frequency differences during normal operation, since the circulating current caused by a frequency or phase difference between inverter outputs does not produce a synchronizing effect, as in the case of synchronous alternators. The voltage waveforms generated by synchronized inverters should also be approximately the same, in order to prevent the circulation of excessive harmonic currents. In the case of load transfer from one inverter to another, the outgoing inverter should be rapidly disconnected before an excessive frequency difference arises.

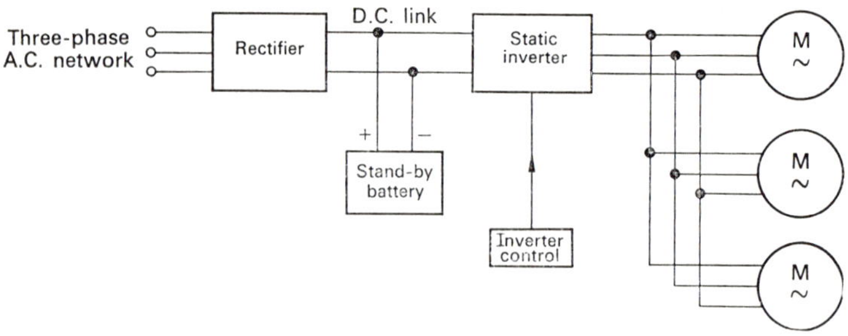

FIG. 7.3. D.C. link converter with stand-by battery for continuity of operation.

7.6. HARMONIC DISTORTION DUE TO STATIC CONVERTERS

In a cycloconverter or d.c. link converter, the current drawn from the a.c. supply network is non-sinusoidal and, therefore, the circuit acts as a generator of harmonic currents. These harmonic currents develop harmonic voltage drops in the network impedance, causing distortion of the supply voltage delivered to the static converter itself and to other consumers in the area.

In general, the harmonic currents on the a.c. side of a rectifier or inverter are of order $k = nP \pm 1$, where P is the pulse number of the converter and n is an integer. The pulse number is the number of thyristors which successively begin to conduct during one cycle of the supply frequency. The six-phase half-wave circuit is therefore classified as a six-pulse system, and so is the three-phase bridge circuit. The lowest order harmonics generated by these circuits are the fifth and the seventh. A twelve-pulse converter is obtained if two six-pulse groups, with a 30° phase displacement, feed a common d.c. load. Similarly, two twelve-pulse circuits with a 15° displacement form a twenty-four-pulse system. Increasing the pulse number reduces the total harmonic content by eliminating low-order harmonics, but correct balance between converter groups is essential. In practice, commutation overlap reduces the amplitudes of the harmonic currents below the theoretical values for instantaneous commutation. Firing delay reduces the overlap angle and so tends to increase the harmonic amplitudes.

Distortion of the supply voltage waveform is objectionable, as it may cause overheating of power-factor-correcting capacitors. At the harmonic frequencies, undesirable series resonance may also occur between the power-factor-correcting capacitance and the transformer reactance. A certain amount of telephone interference is also possible, and lighting flicker may occur. In order to avoid these difficulties, supply authorities specify that the converter rating must not form an excessive part of the total load on the system. The Electricity Council of Great Britain has drawn up Recommendation G.5/2[(160)] which specifies the harmonic current limit in terms of the supply voltage. The recommended limits were obtained from theoretical calculations supplemented by actual measurements. The total permissible harmonic currents on a 415 V distribution system are specified as: 65 A for the fifth harmonic; 42 A for the seventh harmonic; 63 A for the eleventh harmonic and 53 A for the thirteenth harmonic. In order to satisfy these limits, the rating of a converter with zero delay angle should be less than 250 kW for a six-pulse converter, and less than 750 kW for a twelve-pulse system. Larger converter ratings are permitted at higher distribution voltages. In large static drives of several hundred horsepower, the pulse number of the converter may have to be increased in order to reduce the harmonic content to an acceptable level. This increases the cost of the static drive and may render it uneconomic. Transformers are necessary to produce the correct phase displacement between converter groups, and this also increases the cost of the installation. Harmonic filters are sometimes used but they are bulky and costly and their effectiveness varies with changes in the a.c. system.

The distortion of the supply voltage due to a converter load may cause interaction between static converters. Thyristor circuits may also generate radio-frequency interference (RFI) which, if it is not suppressed, is propagated through the supply system and sometimes causes interaction between neighbouring converter installations. The electromagnetic radiation is generated by the sudden changes of current which occur in thyristor circuits. The electromagnetic interference is partly radiated directly but is mainly propagated in the supply lines. It may cause interference to broadcast reception, and is also liable to cause interaction between thyristor installations by coupling transient e.m.f.s into sensitive trigger circuits, or by the introduction of excessive dv/dt transients into anode circuits. The danger of interaction may usually be eliminated by good circuit layout, by shielding sensitive control circuits and by fitting dv/dt suppression circuits.

7.7. THE ECONOMICS OF STATIC VARIABLE-FREQUENCY A.C. DRIVES

In recent years, the d.c. motor powered by a thyristor converter has been widely adopted as a variable-speed drive. However, the static adjustable-frequency a.c. drive is becoming increasingly competitive as thyristor costs continue to fall with increasing production and improved manufacturing techniques.[5] Nevertheless, as explained in Section 1.2, the static a.c. drive is quite expensive, but the cost is readily justified, even in the case of single-motor drives, when adverse ambient conditions exist, or the demanded speed accuracy cannot be satisfied by a conventional variable-speed drive.

A brushless a.c. motor is ideal for applications requiring a minimum of routine maintenance and inspection, and the static a.c. drive is being applied in difficult environmental conditions where the regular maintenance of the d.c. machine is unacceptable. Operation in explosive or radioactive atmospheres, and underwater applications, are typical situations where the capital cost of a static a.c. drive is not difficult to justify. The robust a.c. motor is also advantageous for traction applications, and an a.c. machine can be fitted in a limited space where a d.c. motor or a.c. commutator motor cannot be accommodated. However, the high accuracy and long-term stability of the a.c. drive also render it attractive under less arduous operating conditions. Since the output frequency and speed are determined by an independent reference oscillator, the speed regulation on load of a synchronous or reluctance motor drive is zero. In addition, moderate fluctuations in network voltage and frequency do not affect the speed of the controlled motors. This standard of performance cannot be achieved with other drives.

It is difficult to make an exact economic comparison between the static variable-frequency a.c. drive and other variable-speed drives, because the economics of thyristor converters are constantly improving, as the cost of thyristor devices decreases and their dynamic characteristics are improved. However, it can be asserted that for multi-motor applications requiring precise speed co-ordination, the static a.c. drive is the most economic solution. By using cheaper a.c. motors in place of d.c. motors, the machine costs are reduced to a minimum, and the resultant saving can be set against the cost of the static frequency converter. When large numbers of machines are involved, the cost of the static frequency converter is completely offset in this manner.

With falling thyristor prices, the single-motor variable-frequency drive is also becoming competitive with the converter-fed d.c. motor. The cost of a cage-rotor induction motor is one-sixth, or less, of the cost of a d.c. motor of the same horsepower rating. At present, the cost of an induction motor in a single-motor variable-frequency drive is usually less than 10 per cent of the overall drive cost but this is partly due to the cheapness of the standard cage-rotor induction motor. It might appear, however, that the complex converter circuitry of a solid-state a.c. motor drive would inevitably cause it to be more expensive than a d.c. drive, but this statement must be qualified. If the environmental conditions are hazardous, or the demanded speed accuracy is high, the provision of a suitable d.c. motor with a conventional closed-loop analogue system of speed control of the required accuracy might prove more expensive.

In spite of the double power conversion involved, the efficiency of the d.c. link converter can be 90 per cent, or more, and the overall efficiency of the static a.c. drive is very similar to that of the converter-fed d.c. motor. This is significantly greater than the efficiency of a conventional Ward–Leonard drive or a variable-frequency a.c. system employing a rotating frequency converter. The static frequency converter has a higher capital cost, particularly when regenerative braking is specified, but the static drive has a better performance, lower maintenance costs and less down-time. It also has reduced energy costs due to its higher efficiency and lower installation costs, since expensive foundations and precise alignment are unnecessary.

For smaller drives, the six-step inverter is satisfactory and economically attractive. When the drive horsepower increases, and available thyristor ratings are insufficient, the twelve-step circuit is used instead of simply connecting extra thyristors in parallel in the six-step circuit. In this manner, the advantages of twelve-step operation are achieved at the expense of some additional control circuitry. Multiple-pulse-width modulation also provides an excellent output waveform but requires complex control circuitry. The harmonic losses in the a.c. motor are usually small, and, if necessary, they can be reduced with special machine designs. It is often more economical, however, to use standard a.c. motors to reduce capital expenditure at the expense of slightly increased running costs. As explained in Section 3.6.4, a variety of voltage control methods is available, and the technique which is adopted in practice is determined by economic considerations and the system performance required.

As thyristor ratings increase, inverter manufacturers can produce correspondingly large inverter kVA ratings.[(33)] Development effort is also directed towards producing static a.c. drives of less than 10 kVA, to compete with d.c. drives in this range. Power transistor ratings have been significantly increased, and they can now compete economically with thyristors for converter ratings of 5 kVA or less. In the future, the excellent performance of the static a.c. drive at all power levels is bound to attract widespread application in many industrial processes, and so provide standards of accuracy and reliability which hitherto were impossible.

Some of the static a.c. drives which are described in the succeeding chapters are significantly cheaper than the static variable-frequency system, but their steady-state and dynamic performance characteristics are not as attractive.[(161)] Consequently, the drives described in Chapters 9 and 10 are being applied in situations where a somewhat lower standard of performance is satisfactory.

CHAPTER 8

Commutatorless D.C. Motors

8.1. INTRODUCTION

In static variable-frequency drives using synchronous motors, the inverter output frequency is usually determined by a high-stability reference oscillator. In this manner, a precise shaft speed is obtained which is independent of load. If the gating of the static inverter is controlled by a position transducer on the rotor shaft, the synchronous motor is endowed with the flexible variable-speed characteristics of the d.c. motor. The static inverter and position transducer can be regarded as a solid-state commutator which performs the functions of the mechanical commutator, giving a variable-speed machine with all the versatility of the d.c. motor but none of the disadvantages of mechanical commutation.[57,162–4] Transistors have been used as the switching elements for small motors,[165–6] but thyristors are necessary for larger powers.

Conventional d.c. armature windings have been used with solid-state commutators which simulate the progressive switching action of the multi-segment mechanical commutator. These machines may be operated as commutatorless d.c. motors by fitting a d.c.-excited field system and controlling the transistor or thyristor switching from the rotor shaft. The armature winding and solid-state commutator establish an air-gap magnetic field which rotates synchronously relative to the armature as in a conventional polyphase a.c. machine. Consequently, any standard form of rotor, such as a squirrel-cage rotor, permanent-magnet rotor or reluctance rotor, may also be fitted.

8.2. SOLID-STATE COMMUTATION

In a conventional d.c. motor, a number of coils are distributed around the armature surface and interconnected to form a closed winding. Stationary poles establish a magnetic field in which the armature rotates. Current is also supplied to the armature through the commutator brushes, and the armature magnetic field is established in the direction of the brush axis. The motor torque is due to the tendency of the armature field to align itself with the main field, and maximum torque is developed when the two fields have a space phase displacement of 90 electrical degrees. The provision of a large number of commutator segments results in a unidirectional armature field which reacts with the main field to produce a steady output torque having negligible ripple. In a solid-state commutator, the number of "segments" is reduced, and this causes the output torque to pulsate about its mean value.

Each armature tapping in a conventional d.c. motor is connected to a commutator segment. When a segment passes under a brush, the corresponding armature tapping is connected to the positive or negative supply terminal. In order to reproduce this behaviour in a solid-state commutator, each commutator segment is replaced by a pair of thyristors. One of these thyristors permits current flow from the positive rail to the armature tapping, and the other thyristor allows current flow from the tapping to the negative rail. The mechanical commutator automatically switches each coil at the correct rotor position, but the gating of the thyristors must be controlled by a position transducer on the rotor shaft. This ensures that the switching frequency is proportional to the rotor speed, and maintains the armature field in space quadrature with the field poles. The position transducer may take the form of a mechanical switch directly coupled to the motor shaft, but electromagnetic, optical or Hall-effect transducers may also be used.[167]

When a mechanical commutator is replaced by its solid-state counterpart, the number of switching operations is reduced in order to limit the number of thyristors and improve their utilization. Figure 8.1 (a) shows a ring winding and an elementary three-segment commutator. Figure 8.2(a) shows the equivalent semiconductor arrangement in which three arma-

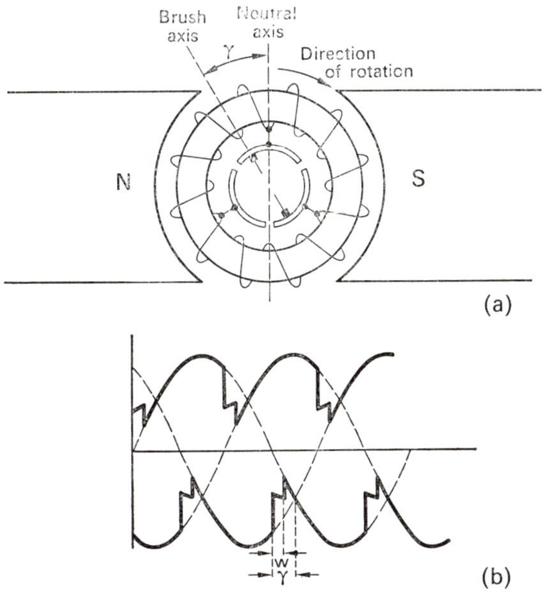

FIG. 8.1. Elementary d.c. motor with three commutator segments: (a) basic construction; (b) brush voltage waveform.

ture tappings are each connected by a pair of thyristors to the positive and negative rails, giving the familiar three-phase bridge circuit. If the induced e.m.f. between segments has a sinusoidal variation with time, the three-phase waveforms of Fig. 8.1 (b) and Fig. 8.2(b) are obtained. These voltage waveforms show the close correspondence between the two systems. In Fig. 8.1 (b) for the mechanical commutator, the brushes are displaced from the neutral axis, and the brush width, w, is taken into consideration. Figure 8.2(b) for the solid-state commutator shows the voltage waveforms for the three-phase bridge circuit operating as an

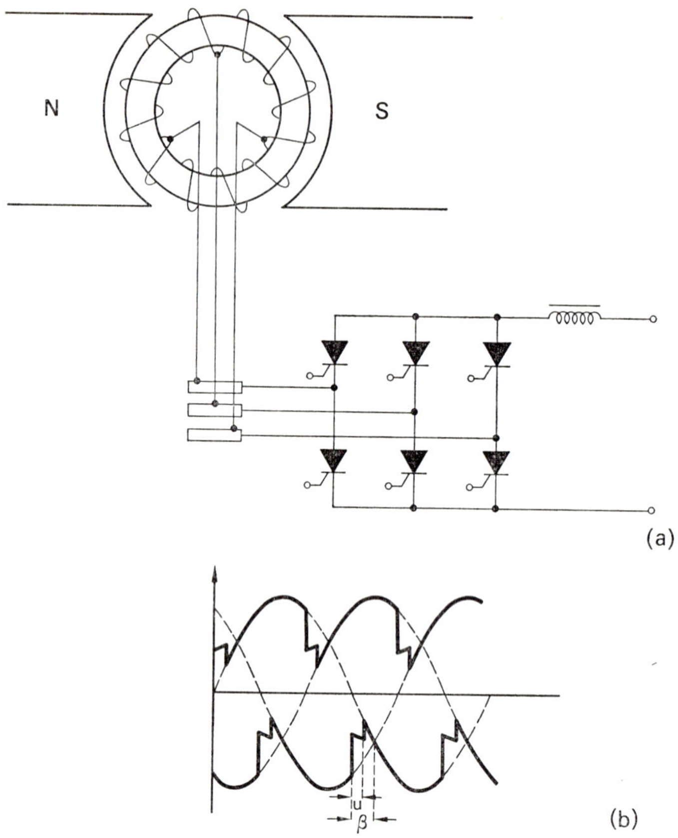

FIG. 8.2. Replacement of mechanical commutator of Fig. 8.1 by a thyristor bridge circuit.

inverter, with the firing point advanced by an angle β to allow sufficient margin for commutation overlap and turn-off.

A large number of commutator segments is required in the conventional d.c. machine, since the voltage between segments is limited by the intersegment insulation. This consideration does not arise in the solid-state commutator, and satisfactory operation is obtained with a limited number of commutations per revolution. The commutator in a conventional d.c. motor is placed on the rotor to avoid the necessity for rotating brushes, but the construction is inverted when a solid-state commutator is employed. A field winding and a pair of slip-rings are then placed on the rotor, and the armature slip-rings of Fig. 8.2(a) are unnecessary. If the armature ring winding is replaced by a phase-wound armature, the machine has the same construction as a three-phase synchronous motor of normal manufacture. Torque pulsations may be troublesome at low speeds, but their effect can be reduced by fitting a substantial damper winding which also improves the steady-state and transient characteristics of the motor.[168-9]

As explained above, a three-phase synchronous motor supplied by a bridge inverter is equivalent to a simple d.c. motor. Thyristors are triggered in the correct sequence by a position transducer on the rotor shaft, and turn-off occurs naturally, as in any phase-controlled inverter connected to a live a.c. supply. In operation, the system behaves like a d.c. motor with a mechanical commutator. The motor speed is proportional to the direct voltage input to the static inverter or solid-state commutator, and speed also varies approximately as the inverse of the field current. The normal synchronous speed of the motor on a fixed-frequency supply is now meaningless, and the upper speed limit is determined by the mechanical limitations of the rotor construction. However, the drive has the disadvantage that the phase-controlled inverter cannot recover after a commutation failure. In addition, the alternating e.m.f.s required for inverter commutation are not available at standstill and low speeds, and auxiliary commutating circuits must be used for these operating conditions.[170-1] Other forms of commutatorless d.c. motor have been constructed using forced-commutation methods at all speeds and, in fact, any synchronous motor fed by a variable-frequency d.c. link converter can be operated as a commutatorless d.c. motor by controlling the inverter frequency from a position transducer on the motor shaft. In Fig. 8.3 the synchronous motor fed by a static inverter can be regarded as a commutatorless d.c. motor. Speed control is obtained by variation of the rectifier delay angle, since this alters the direct voltage delivered to the solid-state commutator.

A cycloconverter, whose output frequency is controlled from the shaft of a synchronous motor, also constitutes a form of commutatorless d.c. motor in which there is no obvious d.c. link. When powered by a d.c. link converter, as in Fig. 8.3, or by a cycloconverter, the synchronous motor is equivalent to a d.c. shunt motor and is described as a shunt commu-

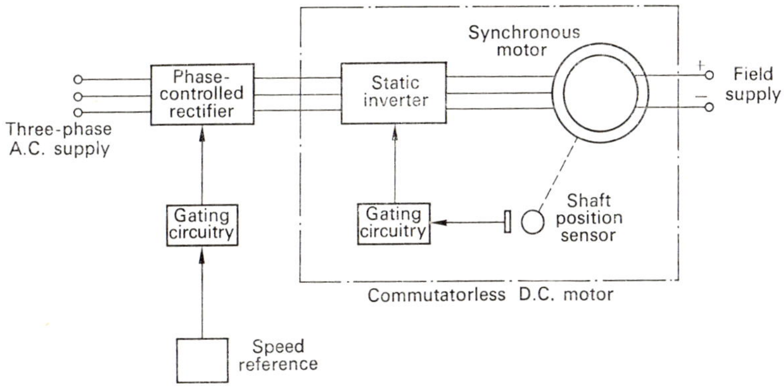

FIG. 8.3. Commutatorless d.c. motor fed by a phase-controlled rectifier.

tatorless motor.[172] A series commutatorless motor, having the operating characteristics of a d.c. series motor, is obtained by supplying the field winding from a bridge rectifier which is in series with the armature winding, or alternatively, a doubly-fed wound-rotor induction motor may be used.[173] A synchronous motor drive of this type was constructed using thyratrons before the development of the thyristor, and was described as a "thyratron motor".[86] A 400-h.p. scheme was in operation for many years in the United States as a power-station auxiliary drive.[87] The increased reliability and efficiency of thyristor circuits has improved the performance of such schemes and they are now a more attractive practical proposition.[174]

8.3. UNCONVENTIONAL COMMUTATORLESS MOTORS

In the commutatorless d.c. motor of the last section, a standard synchronous motor is operated at variable speed by connecting it to an a.c. supply whose frequency is proportional to the shaft speed. The static inverter or cycloconverter generates a near-sinusoidal voltage waveform and hence the motor operation does not differ greatly from normal operation on an a.c. mains supply. However, it is not essential to use standard a.c. motors and sinusoidal waveforms and better machine utilization may be possible using unconventional forms of motor construction and non-sinusoidal supplies.

In the conventional d.c. motor, a multi-segment commutator produces a rectangular distribution of armature current which reacts with the main field flux to produce the electromagnetic torque. Several attempts have been made to simulate statically the progressive switching action of the multi-segment mechanical commutator.[175-9] These special machines are obviously more expensive than standard designs and cannot be justified economically unless the new design offers some advantages compared with conventional forms of construction. As yet, these special machines have not proved sufficiently attractive to merit their adoption by industry, but continued research may result in the development of an improved variable-speed motor.

8.3.1. *D.C. Armature Windings with Solid-state Commutators*

As shown in Section 8.2, a synchronous motor supplied by a three-phase bridge inverter is equivalent to an armature winding with an elementary three-segment commutator. Each commutator segment is replaced by a pair of thyristors, and the armature winding and bridge circuit may be redrawn as shown in Fig. 8.4(a). The number of "segments" is readily increased by introducing additional tappings symmetrically spaced around the ring winding. Thus, the six-phase bridge circuit in Fig. 8.4(b) is equivalent to a six-segment commutator. The armature winding and its thyristor circuitry are naturally placed on the stator, and the d.c.-excited field system is mounted on the rotor. The armature winding and solid-state commutator produce a rotating air-gap field whose speed and direction of rotation can be controlled. As already explained, an inverted d.c. motor with a stationary armature is obtained when the thyristor switching frequency is governed by a position transducer on the rotor shaft. The armature m.m.f. wave then rotates at the same speed as the main field flux and is always in quadrature with it. However, operation is not confined to the commutatorless d.c. mode. If the thyristor switching frequency is determined by an independent variable-frequency oscillator, the d.c.-excited rotor locks into synchronism with the rotating field, and a variable-speed synchronous motor is obtained. Alternatively, a squirrel-cage rotor may be fitted to give a variable-speed induction motor.

Conventional d.c. armature windings may be adapted for solid-state commutation by connecting the armature tappings to pairs of thyristors rather than commutator segments. Figure 8.5 shows a simple lap winding with a solid-state commutator and, when TH1 and TH4′ are gated, a current can circulate through the lap winding as shown, producing the usual triangular armature m.m.f. distribution. The m.m.f. wave steps forward by one coil pitch when TH1 and TH4′ are turned off and TH2 and TH5′ are gated on. Hence, the armature m.m.f. wave does not rotate uniformly but moves in discrete steps. A large number of armature coils and thyristors will give a smoother field rotation and so reduce the rotor losses and undesirable torque pulsations. On the other hand, as the number of thyristors is increased, the conduction periods become shorter and the device utilization is poorer.

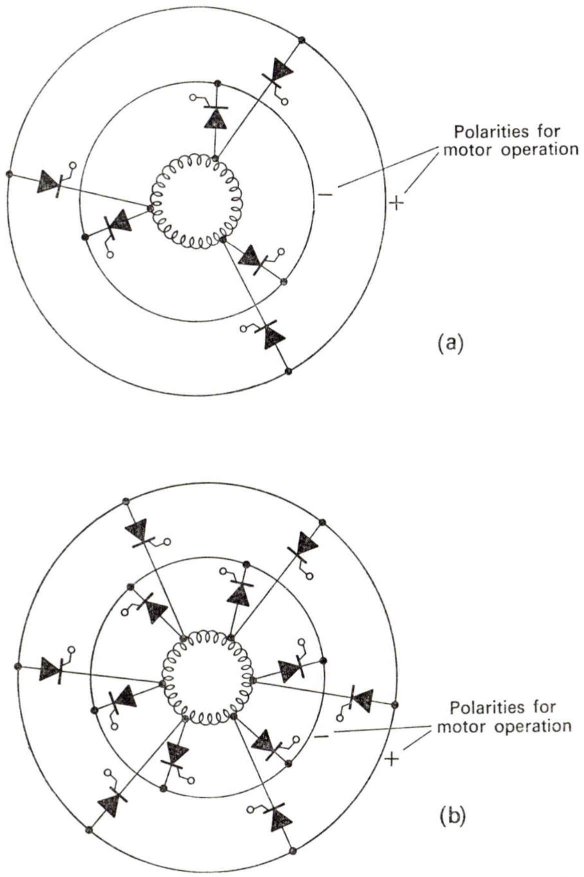

FIG. 8.4. Armature ring winding and solid-state commutator: (a) "three-segment" commutator; (b) "six-segment" commutator.

Individual armature coils may also be controlled by series thyristor switches as shown in Fig. 8.6. A number of adjacent coils are normally allowed to conduct simultaneously but, nevertheless, the coil utilization is not as good as in the lap winding. The armature m.m.f. wave is again stepped forward by sequential switching of the coils.

When these solid-state commutators are supplied with a direct voltage, a forced-commutation technique is necessary to turn off a conducting thyristor. If, however, the solid-state commutator has an a.c. supply, the thyristor current has a half-wave rectified waveform and each thyristor automatically regains control at the end of a conducting half-cycle. This technique permits considerable simplification in the control and turn-off circuitry and is used by Lamb in a 50-Hz single-phase a.c. motor.[175] The parallel winding arrangement of

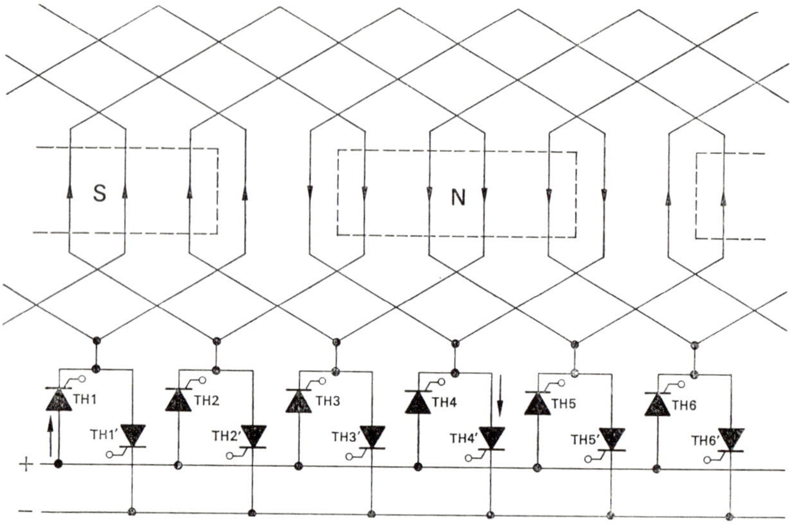

FIG. 8.5. Armature lap winding and solid-state commutator.

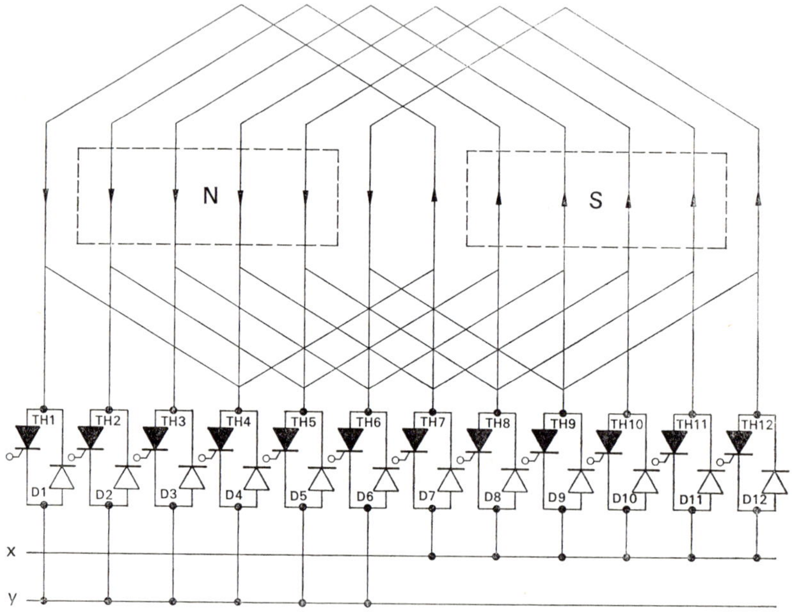

FIG. 8.6. Individual armature coils controlled by a solid-state commutator.

Fig. 8.6 is employed and, when TH1 is gated, a half-wave rectified current will circulate in the coil from line X to line Y through TH1 and diode D7. Electromagnetic torque is developed by the interaction between the current-carrying conductors and the main field flux. When the armature conductors move into the influence of an adjacent pole of opposite polarity, the gating signal for TH1 is removed and it turns off. Thyristor TH7 is then gated to allow current flow in the reverse direction, and the other coils are similarly switched on and off in sequence. This is equivalent to maintaining a space phase angle of 90° between the armature m.m.f. axis and the main field axis and, consequently, a unidirectional torque is developed.

The machine operates as an inverted d.c. motor at low speeds, and each coil conducts for several half-cycles during the passage of a rotor pole across it. The number of conducting half-cycles decreases as the rotor speed increases. If the rotor speed is excessive, the half-cycle of armature current continues after the conductors have come under the influence of the adjacent pole and a braking torque is produced. Also, when the conductor nears the quadrature axis, the induced e.m.f. due to the main field is small, and excessive coil current may result. Higher rotor speeds are possible when the machine is operating in the synchronous or induction motor modes.

Thyristor triggering was originally controlled by a mechanical switching device consisting of two switching rings which were driven by the motor shaft or by a small independent variable-speed motor. Subsequently, a completely static switching unit was developed.[180]

In a more recent paper,[181] Lamb has described an improved commutatorless motor which is suitable for large outputs. The solid-state commutator is normally fed from a three-wire d.c. supply, but an a.c. source is necessary for starting and low-speed running. When operating on the d.c. supply, thyristor commutation is achieved by means of induced armature e.m.f.s which are generated by auxiliary windings in the field system.

8.3.2. *Forced Commutation by an Auxiliary E.M.F.*

When the solid-state commutator has a d.c. supply, forced commutation may be achieved by means of two auxiliary d.c. supplies, as suggested by Tustin.[176] This circuit employs a series armature winding, and Fig. 8.7 shows a ring winding and thyristor commutator which can establish a two-pole field and move it in either direction in 60° steps. The main

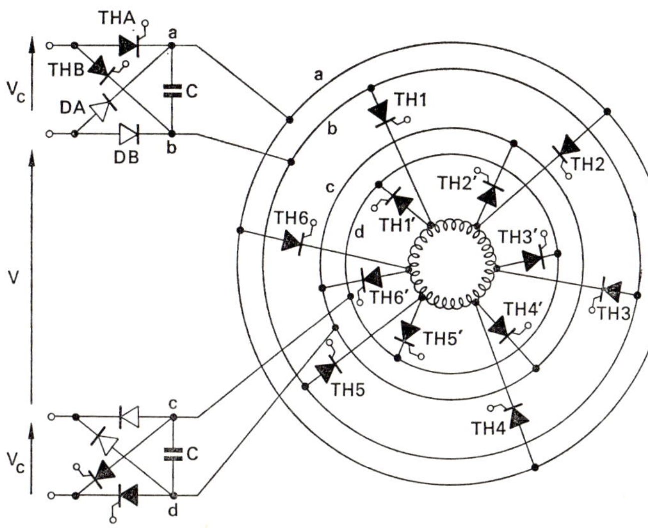

FIG. 8.7. Solid-state commutator with auxiliary commutating circuits.

d.c. supply voltage is V, and the switching frequency is determined by the two commutating voltages, V_c. By means of an auxiliary switching circuit, the direct voltage, V_c, can be applied in alternate directions between the two positive rails, a and b. Triggering THA, for instance, raises rail a to a potential $(V + V_c)$, while rail b is held at potential V, through diode DB. The polarities are reversed by firing THB, and the commutating capacitor C ensures that THA and THB never conduct simultaneously. A similar switching operation can be performed between the two negative rails, c and d.

In the two-pole arrangement of Fig. 8.7, the current must enter at a tapping on the ring winding and leave at a diametrically opposite point. The commutation process causes a progressive rotation of the current axis, and commutations at the positive and negative rails are synchronized in order to preserve the correct field pattern. During commutation, the voltage V_c reverses the current in the winding segment between two tappings, but the current in the rest of the winding is uninterrupted.

Each tapping point on the ring winding is connected by a pair of thyristors to the positive and negative supply lines. As already shown in Section 8.2, each thyristor-pair is equivalent to one segment of a mechanical commutator, and the winding arrangement of Fig. 8.7 therefore corresponds to a six-segment commutator. A comparison with Fig. 8.4(b) emphasizes that the circuit is essentially a six-phase bridge with the positive and negative lines split into two rails having a commutating circuit connected between them. Thyristor anodes are connected alternately to positive rails a and b, so that the anode of the incoming thyristor can always be made more positive than the conducting anode by an amount V_c. Similarly the thyristor cathodes are connected alternately to the negative rails c and d.

Assume TH1 is delivering current to the armature via diode DB and the positive rail b, while TH4′ returns the current to the negative side of the supply. For simplicity, the commutation process on the positive side will be examined separately. The transfer of current from TH1 to TH2 is initiated by firing THA and TH2 simultaneously. This raises rail a to a potential $(V + V_c)$, while rail b is held at potential V by the conduction of DB. The commutating voltage, V_c, increases the current in TH2 and reduces the current in TH1 to zero. This reverses the current in the section of winding between TH1 and TH2, and also charges capacitor C to a potential V_c, with the upper plate positive. The next commutation is initiated as soon as TH1 and TH4′ are blocked. TH3 and THB are then triggered, and capacitor C discharges through THA and THB, reducing the current in THA to zero and turning it off. TH2 now conducts from rail a through diode DA, while THB raises rail b and the anode of TH3 to a potential $(V + V_c)$. Commutation can now proceed from TH2 to TH3, and so conduction progresses from thyristor to thyristor, and the armature m.m.f. wave is stepped forward. Each commutation is initiated as soon as the preceding commutations on the positive and negative sides are completed. The rising blocking voltages on the outgoing thyristors are used to generate the firing pulses which initiate the next pair of commutations.

Tustin's circuit is analogous to a conventional d.c. machine with interpoles. These are auxiliary poles placed midway between the main field poles in order to improve commutation. In the brief period during which two commutator segments are short-circuited by a brush, the current in the shorted coil must be reversed. However, the self-induced e.m.f. in the coil opposes this rapid current change, and an externally induced e.m.f. is necessary to cancel the self-induced e.m.f. The magnitude of this commutating e.m.f. must increase as the available commutation time is reduced and as the coil current is increased. These conditions are satisfied in a conventional d.c. machine when the interpoles are excited by windings in series with the armature. If there is no magnetic saturation, this gives a commutating e.m.f. which is proportional to armature current and speed. In Tustin's circuit, a commutating e.m.f. V_c is also provided to oppose the self-induced e.m.f. and reverse the current in the coil undergoing commutation. The magnitude of V_c determines the commutating time and so governs the speed of rotation of the armature field. An increase in armature current increases the commutating time and reduces the angular speed of the field.

CHAPTER 9

Induction Motor Speed Control by Variation of the Stator Voltage

9.1. INTRODUCTION

Most a.c. motors run at a speed which is determined principally by the frequency of the a.c. supply. The speed of a synchronous motor is completely determined by the stator frequency, but the induction motor runs asynchronously at a speed which is normally within 5 per cent of the synchronous speed. On a fixed-frequency supply, the torque developed by the induction motor, at a given slip, varies approximately as the square of the applied voltage, and steady-state operation occurs when the motor torque balances the load torque. Consequently, the rotor slip is determined by the load torque and the applied voltage, and continuous speed control may be obtained by stepless adjustment of the stator voltage without any alteration in the stator frequency.

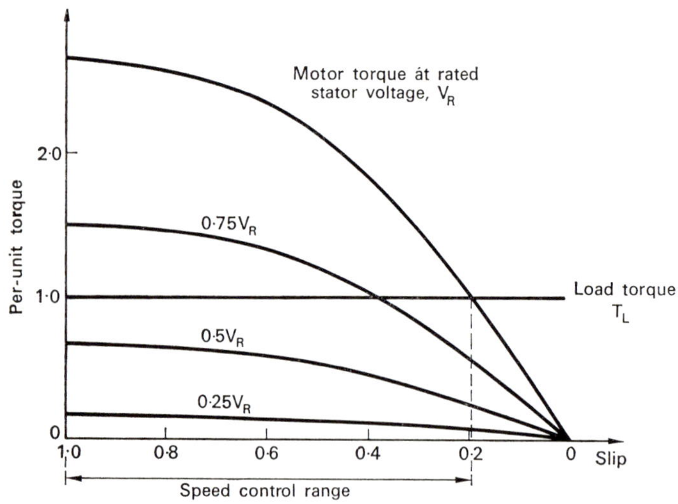

FIG. 9.1. Speed control of a constant-torque load by stator voltage control.

As the stator voltage is reduced, the rotor speed decreases, but the maximum torque available from the motor also decreases, and the speed range is limited with a constant-torque load. A high-resistance rotor modifies the torque-speed characteristics, and wide-range speed control is possible (Fig. 9.1), but the resistance losses are excessive and the operating efficiency is low, particularly at reduced speeds. The low-speed performance is poor because the motor current, at a given slip, is proportional to the applied voltage, whereas the electromagnetic torque varies approximately as the square of the voltage. Consequently, the torque per ampere is lower at reduced speeds, and large currents are required to develop appreciable torque. However, in fan or pump drives, the load torque varies approximately as the square of the speed. Hence, the torque required for starting and low-speed running is small, and may be obtained, without excessive overheating, from a voltage-controlled induction motor with a normal full-load slip of about 10 per cent (Fig. 9.2).

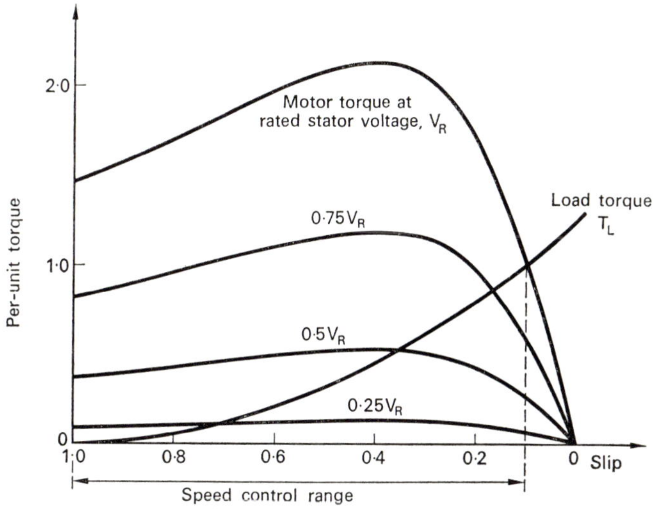

FIG. 9.2. Speed control of a fan load by stator voltage control.

The stator voltage is reduced by connecting variable external impedances between the stator terminals and the a.c. supply mains. Saturable reactors have been used in the past to perform this function, but thyristor circuits now offer several advantages and are competitively priced.[182–3] The thyristor unit is more compact, despite the heat sink requirements, and it weighs considerably less. It also has a higher efficiency, and the response time is a half-cycle of the supply frequency, compared with a time lag of 0·1 second, or more, for a saturable reactor circuit. In addition, thyristor units from different manufacturers are interchangeable, whereas saturable reactor characteristics vary considerably. The basic thyristor circuit which is employed is the series a.c. controller, already described in Section 2.4.2. This consists of two thyristors connected in inverse parallel, or "back-to-back", and triggered symmetrically at identical points in their anode-to-cathode voltage cycles. By inserting these circuits in the stator leads, and controlling the conduction periods of the thyristors, the effective voltage delivered to the motor can be varied from zero to the full supply voltage.[184–8] The motor is subjected to a chopped sine-wave voltage and the supply currents also have a high harmonic content, but satisfactory operation has been achieved with small and medium-sized induction motors up to 100 h.p., or more.

Stator voltage control eliminates the complex circuitry of the variable-frequency schemes discussed in earlier chapters and, consequently, is cheaper to install. However, the operating efficiency is poor, and de-rating is necessary at low speeds to avoid overheating due to excessive current and reduced ventilation. Stator voltage control is widely used for fractional horsepower drives, and also for a.c.-powered cranes and hoists where large torque at high slip is only demanded for intermittent portions of the duty cycle. Variable-frequency drives are unnecessarily complex and expensive for such applications.

9.2. THEORETICAL PRINCIPLES OF STATOR VOLTAGE CONTROL

Assume that the motor voltages and currents are balanced and sinusoidal, and also regard the rotoı resistance as constant and independent of rotor frequency. If ω_1 denotes the synchronous angular velocity and ω is the rotational angular velocity, the fractional slip is defined by $s = (\omega_1 - \omega)/\omega_1$. The gross mechanical power output, including windage and friction losses, is $P_{\text{mech}} = T\omega$, where T is the internal motor torque. In accordance with standard theory (see Appendix A), the total power input to the rotor, across the air gap from the stator, is $P_{\text{ag}} = T\omega_1$. The difference between the power input to the rotor and the mechanical power output is dissipated as heat losses in the rotor resistance. Thus, the rotor copper loss is given by

$$P_2 = T(\omega_1 - \omega) = sT\omega_1 = sP_{\text{ag}}. \tag{9.1}$$

The basic power division process in the rotor circuit is now evident. A fraction s of the rotor input power is dissipated as heat, and the remainder, $(1 - s)$, is available as mechanical power output. Hence

$$P_{\text{mech}} = (1 - s)\,P_{\text{ag}}. \tag{9.2}$$

Speed control using an adjustable-voltage, fixed-frequency supply is therefore inherently inefficient, since the rotor copper loss increases linearly with slip for a constant motor torque. Equation (9.1) also shows that the rotor copper loss is proportional to the product of torque and slip, and the motor heat generated in a given duty cycle may be quickly approximated by evaluating the product of torque, slip, and time, for each portion of the duty cycle.

The stator current, I_1, contains a load component of current which is proportional to I_2, and also a magnetizing component. At reduced voltages, the magnetizing current may be neglected, and hence the currents I_1 and I_2 are proportional. The large rotor currents corresponding to high-torque, high-slip operation will produce correspondingly large stator currents which may cause serious overheating and damage to the sensitive stator winding. Consequently, the motor must be de-rated for low-speed operation.

Equation (9.1) may be rewritten to give an expression for the motor torque. Thus,

$$T = \frac{m_1 I_2^2 R_2}{s\omega_1}$$

where m_1 is the number of stator phases. Therefore

$$T \propto \frac{I_2^2 R_2}{s} \propto \frac{I_1^2 R_2}{s}. \tag{9.3}$$

Stable operation occurs when the motor torque, T, equals the load torque, T_L. For a fan load, $T_L \propto (\omega)^2 \propto (1 - s)^2$, and equating torques gives the result

$$I_1 \propto I_2 \propto (1 - s)\sqrt{(s/R_2)}. \tag{9.4}$$

This result indicates that the machine currents are inversely proportional to $\sqrt{R_2}$ and, by differentiation, it is found that the currents have peak values when $s = \frac{1}{3}$, that is, when $\omega = \frac{2}{3}\omega_1$.

For a constant-torque load, T_L is constant, and equating motor and load torques gives the result $I_1 \propto I_2 \propto \sqrt{(s/R_2)}$. The stator and rotor currents are again inversely proportional to $\sqrt{R_2}$, but they now increase steadily as the motor speed is reduced to zero, and are even larger in the plugging region where the slip is greater than unity.

These results show that the rotor resistance is an important factor in determining the increase in motor current. If the normal full-load slip of a squirrel-cage motor is small, indicating the presence of a low-resistance rotor, a very large increase in stator current will occur at low speeds when stator voltage control is employed. For satisfactory low-speed operation in a fan drive, the rotor should have a high resistance corresponding to a full-load slip of 10 to 12 per cent.[186] In a constant-torque drive, the rotor resistance must be even greater. Squirrel-cage rotors with variable-resistance characteristics may also be used. The usual deep-bar or double-cage rotor is not very effective, since the eddy-current loss, on which the operation depends, is proportional to the square of the slip, and hence is not significant unless the slip exceeds about 0·5. An improved performance can be obtained by using a solid-iron rotor,[189] or by placing a bar of permanent-magnet material, such as Alnico, in the top of each rotor slot. The hysteresis loss in the Alnico causes an increase in the effective rotor resistance, thereby increasing the torque per ampere at low speeds and giving the motor a better torque-speed characteristic.[190]

Standard wound-rotor induction motors are commonly used in variable-voltage speed-control systems. External rotor resistance is added to modify the torque-speed characteristic so that the torque per ampere at low speeds is increased. Motor heating is reduced, since part of the rotor circuit loss is now dissipated externally. A further improvement in low-speed performance may be obtained by connecting Alnico-cored saturable reactors, known as "Saturistors", in the rotor circuit.[191]

Speed control may also be obtained by deliberately unbalancing the stator terminal voltages. By the theory of symmetrical components, the unbalanced three-phase voltages can be resolved into positive and negative-sequence systems. The positive-sequence set establishes balanced currents of the same sequence and these produce a forward-rotating field which develops forward rotor torque in the usual manner. The negative-sequence voltages produce negative-sequence currents which establish a backward-rotating field and develop a counter-torque opposing motion. The positive and negative-sequence currents both contribute to the motor heating, but the resultant torque is the difference between the positive and negative-sequence torques. The low-speed performance is, therefore, even poorer than that obtained with a symmetrical reduction of the stator voltages, and further de-rating is necessary to limit the temperature rise.[192]

When thyristor switching circuits are used to reduce the stator voltage, either symmetrically or asymmetrically, the resulting voltage and current waveforms are highly distorted, and the additional harmonic losses will further accentuate the overheating problem at low speeds. An exact analysis of induction motor behaviour under such operating conditions is rather difficult and requires an analogue simulation or a numerical solution on a digital computer.[193-8]

9.3. PRACTICAL CIRCUITS[184,186,199]

As already explained in Section 2.4, the effective load voltage in a single-phase a.c. circuit can be varied by a thyristor controller consisting of a pair of anti-parallel thyristors in series with the load. A three-phase version of this circuit is obtained by combining three single-phase circuits in a delta-connection, as in Fig. 9.3. Each pair of back-to-back thyristors controls the voltage delivered to one phase of the stator, and the phase voltage waveform

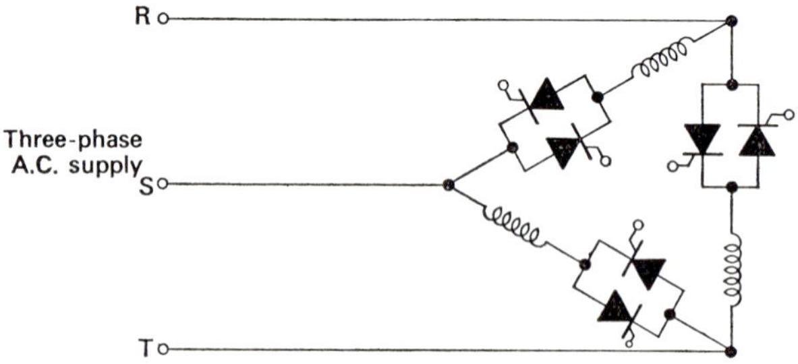

FIG. 9.3. Stator voltage control of a delta-connected induction motor by a series back-to-back thyristor pair in each phase.

consists of a series of sine-wave segments. A phase displacement of 120° is maintained between the sets of gating pulses delivered to each controller in order to produce a symmetrical reduction of the three-phase voltages. In the full-voltage condition, each thyristor receives a gating signal at the start of the positive half-cycle of its anode-to-cathode voltage. The series thyristor pair is then virtually a short-circuit and the stator phase receives a complete cycle of supply voltage. As the thyristor firing is delayed, the conduction period of each thyristor is shortened and the effective stator voltage is reduced. When the firing delay is 180°, the thyristor controller is open-circuited and the motor voltage and current are both zero. Using a switching device in this manner as a non-linear series impedance has the advantage that the power dissipation in the controller is considerably less than the load power, but the switching action also distorts the stator voltages.

Several alternative power circuits are possible. Thyristor controllers may be inserted in the three a.c. lines of a star or delta-connected load, as in Fig. 9.4(a) and (b). For economy, one thyristor in each back-to-back circuit can be replaced by a rectifier diode. The number of semiconductor devices can be reduced if it is decided to operate with unbalanced stator voltages. Back-to-back thyristors can be inserted in two of the three-phase lines, as in Fig. 9.4(c), with the third line connected directly to the supply. Alternatively, the motor can be operated with a thyristor controller in only one of the three-phase lines. However, these unbalanced circuits tend to accentuate the heating problem since some phases are heavily overloaded.

Speed control may be obtained by decreasing or unbalancing the stator voltages of the induction motor, but the direction of rotation can only be reversed by changing the phase sequence of the applied voltages. This is often achieved by means of a mechanical contactor which interchanges two stator leads. Contactor operation takes place after the stator current has been interrupted by removal of the thyristor gating pulses. Consequently, sparkless operation is obtained and the contactor requires little maintenance. However, the stator

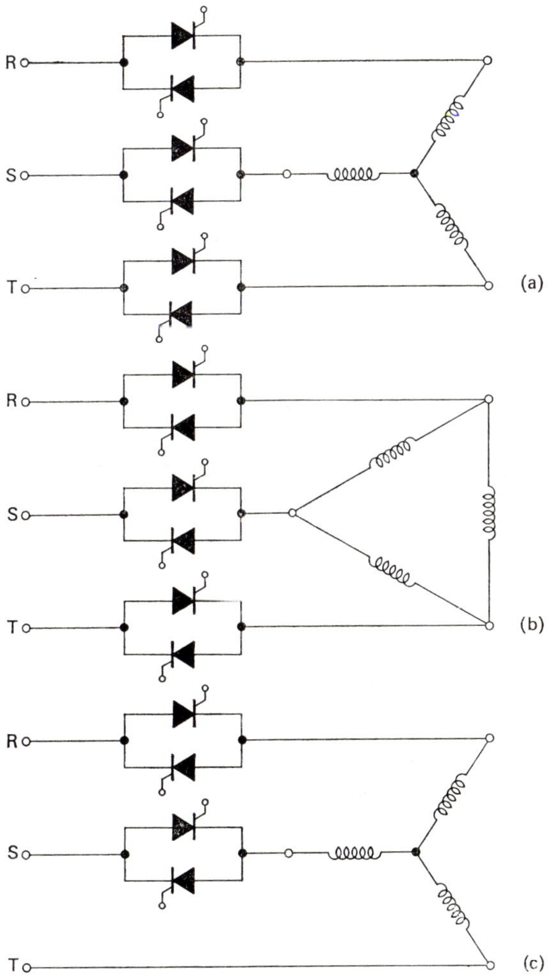

FIG. 9.4. Alternative connections for stator voltage control of a three-phase induction motor.

phase sequence can also be reversed statically by introducing additional thyristor controllers, as in Fig. 9.5. The extra units X and Y are not gated for normal forward rotation of the motor, but only come into operation when A and C are on open-circuit. This reverses the phase sequence of the applied voltages and thereby changes the direction of rotation. It is, of course, essential that X and Y should not be gated until the gating signals for A and C have been removed, or otherwise a short-circuit condition would result. A current detector in series with A and C inhibits firing of X and Y until a zero-current signal is obtained. Similarly, the triggering of A and C is inhibited until X and Y have been open-circuited.

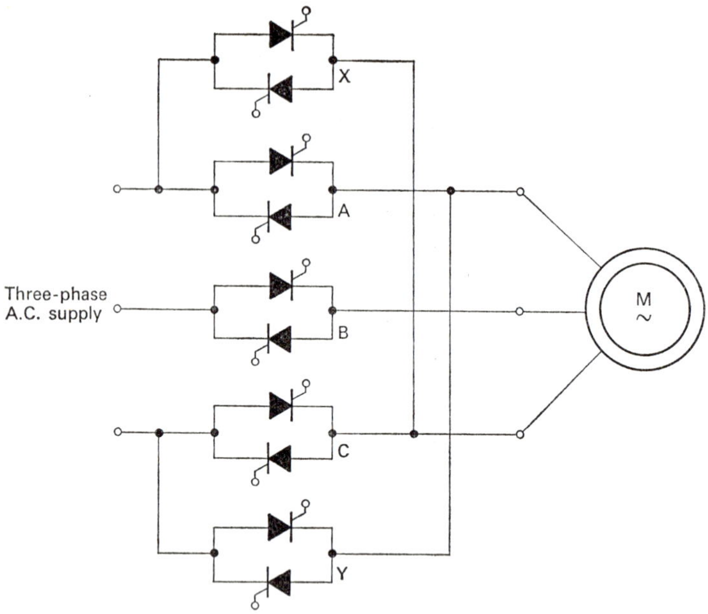

FIG. 9.5. Thyristor configuration for a reversible variable-speed induction motor drive.

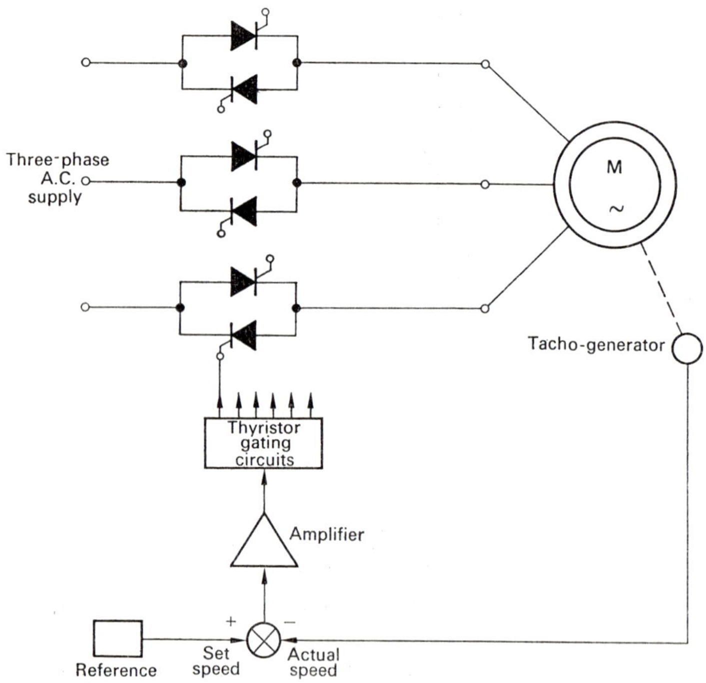

FIG. 9.6. Closed-loop system for induction motor speed control by stator voltage control.

Variable-speed drives using stator voltage control are normally closed-loop systems, as in Fig. 9.6.[200] The d.c. tacho-generator develops a voltage proportional to the motor speed, and this is compared with the d.c. reference voltage representing the desired speed. The difference between the two signals is the error voltage which controls the thyristor firing angles and thereby alters the terminal voltage and motor speed so that the error is reduced. If the reference voltage is greater than the tacho-generator voltage, the conduction periods of the thyristors are increased. The increased stator voltage allows the development of an increased motor torque and hence the speed rises. If the tacho-generator voltage exceeds the reference voltage, the conduction periods are reduced and the motor torque decreases, causing a reduction in shaft speed. When the motor speed is equal to the desired speed, the conduction periods are just sufficient to allow the development of a motor torque equal to the load torque. In a high-gain feedback system, the desired speed can be accurately maintained and there is no necessity for the motor to have a flat speed-torque characteristic, since the output speed is determined by the reference signal rather than the open-loop characteristic of the motor. Stable operation may be obtained at any point of the induction motor speed-torque characteristic.

In the static reversing drive of Fig. 9.5, the polarity of the error signal determines whether motoring or braking torque is developed, and operation is possible in the speed-torque quadrants shown (Fig. 9.7). If the reference voltage is positive and exceeds the tacho-generator voltage, forward torque is produced by gating thyristor controllers A and C, and hence the motor accelerates. If the reference voltage is suddenly reduced, the polarity of the error signal changes, and thyristor units A and C are turned off and X and Y are gated on. This reverses the phase sequence of the stator voltage and the motor operates in the plugging region with a rapid reduction in speed. In a crane or hoist application, when the motor is

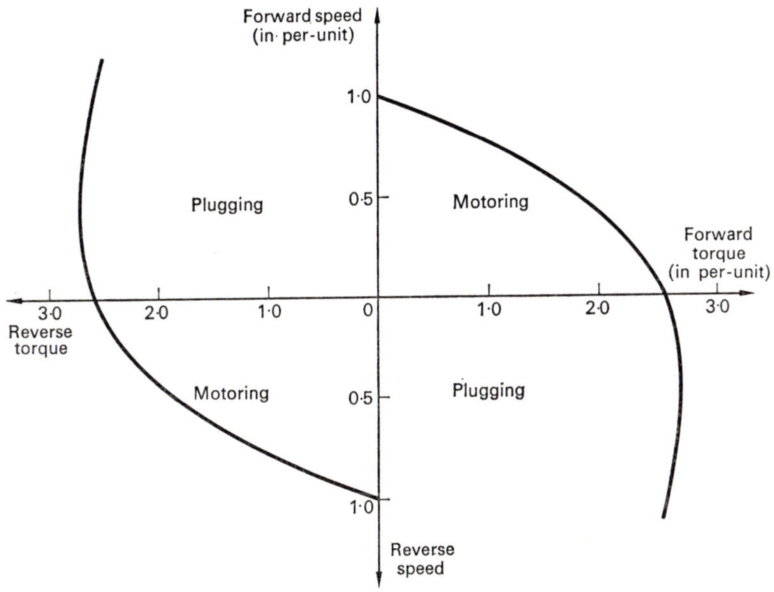

FIG. 9.7. Speed-torque characteristics for a reversible induction motor drive.

lowering an overhauling load, the speed signal is also greater than the reference voltage, and a braking counter-torque is developed which permits steady-state lowering of the load. Regenerative braking in the induction generator region is possible for full-speed lowering of an overhauling load.

9.4. INTEGRAL-CYCLE CONTROL

In the voltage-control methods described above, the effective stator voltage is varied by means of back-to-back thyristor circuits which conduct for portion of each cycle. As an alternative, integral-cycle or time-ratio control may be employed, using the back-to-back circuits as static contactors and alternately switching on and off the supply voltage for a number of cycles at a time.(201) The effective stator voltage is varied by controlling the ratio of on-time to off-time, and speed fluctuations are avoided if the conducting and non-conducting intervals are short compared with the mechanical time constant of the system.(202) Less mains voltage distortion is produced, and less radio-frequency interference is propagated, when using integral-cycle systems rather than the phase-controlled regulators described in this chapter. On the other hand, speed fluctuations may be objectionable, and stepless speed control is not possible when using integral-cycle techniques.

CHAPTER 10

Induction Motor Speed Control by a Static Converter Cascade

10.1. INTRODUCTION

The wound-rotor induction motor is not as rugged as the squirrel-cage motor and cannot operate at such high speeds. However, in a slip-ring machine there is little wear or sparking at the brushes, and hence the maintenance requirements are considerably less than in a d.c. motor. Consequently, the slip-ring induction motor is an acceptable alternative to the squirrel-cage machine in many industrial applications, and the availability of the rotor terminals introduces new possibilities for wide-range speed control. The most important static method of speed control involves the removal of rotor slip power by a converter cascade circuit. The principle of slip-energy recovery offers high efficiency and high reliability and the drive has good control characteristics. Static converter cascades have found considerable acceptance in recent years, particularly for a limited sub-synchronous range of speed variation, when the converter rating can be significantly less than the motor rating.[203-4] Sub-synchronous static converter cascades have been constructed for motor output ratings of up to 8 MW, and drive efficiencies of over 95 per cent are obtained in the megawatt range.[205]

10.2. THEORETICAL PRINCIPLES OF SLIP-ENERGY RECOVERY

The division of power in the rotor circuit has already been studied in Section 9.2. The power delivered to the rotor across the air gap, P_{ag}, is divided between the mechanical power output, P_{mech}, and the rotor copper loss, P_2. The following relations have been derived:

$$P_2 = sP_{ag} \tag{9.1}$$

and

$$P_{mech} = (1 - s)\,P_{ag}. \tag{9.2}$$

Also

$$P_{ag} = T\omega_1$$

where T is the electromagnetic torque developed by the motor, and ω_1 is the synchronous angular velocity.

From these equations, it is obvious that the air-gap power is nearly constant when the induction motor drives a constant-torque load, and hence the rotor copper loss is proportional to slip. Speed control of a wound-rotor motor by the introduction of external rotor resistance is, therefore, inherently inefficient. At half synchronous speed, the air-gap power is divided equally between mechanical power output and rotor I^2R loss, giving an overall efficiency, when stator losses are taken into consideration, of less than 50 per cent. In general at slip s, mechanical power is obtained from the air-gap power with a per-unit conversion efficiency of $(1 - s)$, and the overall motor efficiency is even lower than this. At low speeds, the air-gap power is almost completely dissipated in the rotor circuit, and the efficiency is very poor. The rotor resistance method of speed control is, therefore, uneconomical except for a very small sub-synchronous speed range. However, it is not essential that the slip power, sP_{ag}, be dissipated in resistance losses, as it can be removed from the rotor circuit and utilized externally, thereby improving the overall efficiency of the system. In these cascade connections, the slip power is either returned to the supply network or is used to drive an auxiliary motor which is mechanically coupled to the induction motor shaft.

The operation of the cascade connection may also be regarded as speed control by e.m.f. injection in the rotor circuit. Assume the motor is operating normally at slip s, and an external voltage is applied at the slip-rings in phase opposition to the rotor e.m.f. The resultant decrease in rotor current causes a reduction in motor torque, since torque is proportional to in-phase rotor current, assuming constant air-gap flux (Appendix A). The motor speed therefore decreases due to the braking action of the load. However, as the slip increases, the rotor e.m.f. and current also increase, and stable operation is obtained at some reduced speed when the motor torque again equals the load torque.

When rotor resistance control is adopted, the slip-frequency voltage drop in the external resistor constitutes the injected voltage, but an external e.m.f. source is equally effective and does not introduce excessive heat loss. The main problem in providing a suitable e.m.f. source is that the frequency of the injected e.m.f. must match the rotor slip frequency at all speeds. Hitherto, mechanical frequency-conversion methods have been employed. In the a.c. commutator motor, the frequency-changing action of the commutator is utilized to match frequencies and give a self-contained variable-speed machine. At higher power levels, a conventional slip-ring induction motor is employed, and auxiliary rotating machines in the cascade circuit perform the frequency-changing function. In the Scherbius system shown in Fig. 10.1 a rotary converter rectifies the slip power, and the rectified output drives a d.c.

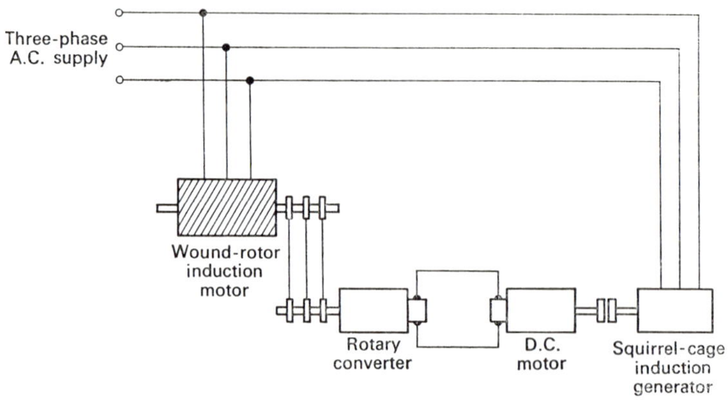

FIG. 10.1. The Scherbius system for speed control of a wound-rotor induction motor.

motor which is mechanically coupled to a squirrel-cage induction generator. The induction generator is driven at super-synchronous speeds and returns the slip power to the a.c. supply. In the Kramer system, the slip power is also rectified by a rotary converter, and the output supplies a d.c. motor which is mechanically coupled to the main induction motor. The slip power is thus converted to mechanical power at the induction motor shaft.

In recent years, static frequency converters have been used to replace the auxiliary machines in the Scherbius system, and this has resulted in a more compact variable-speed drive with an improved operating efficiency and a better dynamic response.[206] The Kramer drive has also been modified by using a diode bridge rectifier in place of the rotary converter, but a d.c. motor is still required to convert the rectified slip power to mechanical power.[207]

10.3. SUB-SYNCHRONOUS STATIC CONVERTER CASCADE

For sub-synchronous speed control, the auxiliary machines of the Scherbius drive are replaced by a static converter, as shown in Fig. 10.2. This consists of a three-phase bridge rectifier which operates at slip frequency and feeds rectified slip power through the smoothing inductor to the thyristor inverter. The inverter returns the rectified slip power to the a.c. supply. The rectifier and inverter are both phase-commutated by the alternating e.m.f.s appearing at the slip-rings and supply busbars, respectively.

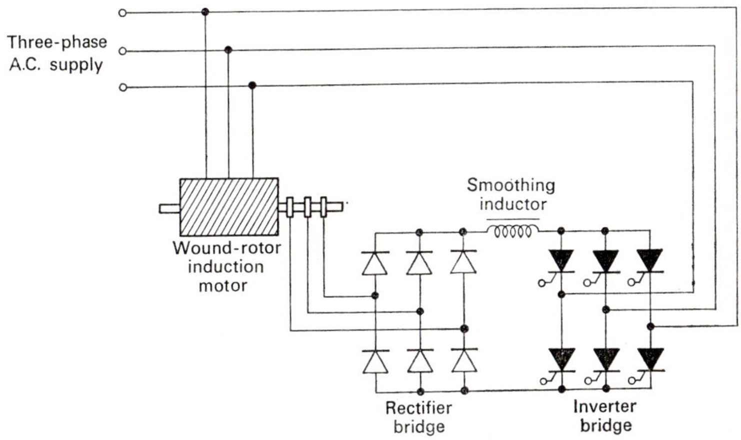

FIG. 10.2. Induction motor and sub-synchronous static converter cascade.

The average counter e.m.f. of the inverter may be regarded as an injected e.m.f. opposing the rectified rotor voltage. If commutation overlap is negligible, the direct voltage output of the uncontrolled three-phase bridge rectifier is obtained from equation (2.7) as

$$V_{do} = 1{\cdot}35 V_r s \qquad (10.1)$$

where V_r is the line-to-line rotor voltage at standstill, and s is the fractional slip.

For a three-phase bridge inverter, the average counter e.m.f. is given by

$$V_d = 1{\cdot}35 V_L \cos \alpha \qquad (10.2)$$

where α is the firing delay ($\alpha \geqslant \pi/2$), and V_L is the a.c. line voltage (Section 2.6).

On no-load, the motor torque is negligible and the rectified rotor current is almost zero. Consequently, the two direct voltages of equations (10.1) and (10.2) must balance.

Thus

$$1{\cdot}35V_r s + 1{\cdot}35V_L \cos\alpha = 0,$$

or

$$s = -(V_L/V_r)\cos\alpha = -a\cos\alpha = a\,|\cos\alpha| \tag{10.3}$$

where a is the effective stator-to-rotor turns ratio of the motor.

Speed control is, therefore, obtained by a simple variation of the inverter firing angle. If a is unity, the no-load speed of the motor can be controlled from near standstill to full speed.

In order to develop load torque, a rotor current I_2 is required, and the rectified rotor voltage must force current flow against the inverter back e.m.f. As the induction motor is loaded, the speed falls slightly and the resulting increase in rectified voltage produces the necessary increase in rotor current.

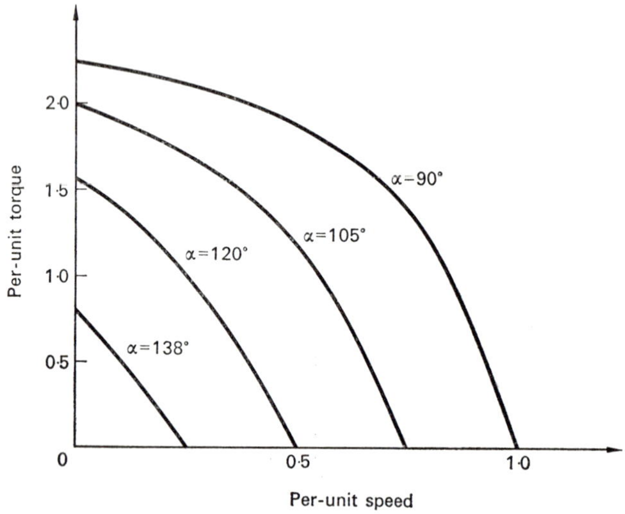

FIG. 10.3. Open-loop torque-speed characteristics for the induction motor and static converter cascade.

If the rotor resistance is small, the fundamental rotor slip power, sP_{ag}, is approximately equal to the d.c. link power.

Thus

$$sP_{ag} = V_d I_d,$$

but

$$P_{ag} = T\omega_1$$

and hence

$$T = \frac{V_d I_d}{s\omega_1}.$$

If the speed droop on load is neglected, equation (10.3) for the no-load slip can be substituted in this result. Substituting also for V_d from equation (10.2) gives the torque expression

$$T = \frac{1{\cdot}35 V_L I_d}{a\omega_1}. \tag{10.4}$$

Thus, the steady-state torque is proportional to the rectified rotor current, I_d, which, in turn, is proportional to the difference between the rectified rotor voltage and the average counter e.m.f. of the inverter. The inverter e.m.f. is constant for a fixed firing angle, and hence the rotor slip increases linearly with load torque, giving a shunt torque-speed characteristic similar to that of a separately excited d.c. motor with armature-voltage control. In practice, the open-loop torque-speed characteristics have the form shown in Fig. 10.3.

10.3.1. *Power Factor*

The principal disadvantage of the sub-synchronous cascade drive is its low fundamental power factor or displacement factor. If the system is designed for wide-range speed control, the full-load power factor at maximum speed may be as low as 0·5, decreasing to 0·3 or less as speed is reduced. This low power factor is partly due to the commutating reactive power which is drawn through the induction motor from stator to rotor by the three-phase bridge rectifier.[208–10] However, the reactive power consumption of the phase-controlled inverter is also responsible for the low power factor of the cascade drive. The average counter e.m.f. of the inverter is a maximum at the lowest controllable speed. Ideally, under these conditions, the firing angle is 180°, but commutation overlap and thyristor recovery time are significant, and hence the inverter firing must be advanced to prevent a commutation failure. This causes the inverter output current to lead the corresponding phase voltage, and the inverter acts as a generator of leading reactive power, that is, a consumer of lagging reactive power (Section 2.5.4). The reactive power consumption increases when the inverter firing point is further advanced to reduce the counter e.m.f. and motor slip. At full speed, the inverter firing angle is 90°, and the inverter kVA is almost completely reactive. At a given firing angle, the net power drawn from the a.c. supply is the difference between the power absorbed by the motor and the power returned by the inverter. However, the total reactive kVA is the sum of the reactive powers absorbed by the motor and inverter, and consequently the system power factor is poor at low speeds when the total active power consumption is small.

As explained in the next section, the cascade drive is frequently operated with a reduced sub-synchronous range of speed control. In order to maintain the system power factor at the highest possible value, the inverter firing angle should be a maximum at the lowest controllable speed. Thus, when the inverter is connected directly to the a.c. supply, the slip-ring voltage at the maximum working slip, s_{max}, should be approximately equal to the a.c. supply voltage. This is achieved by making the stator-to-rotor transformation ratio, a, equal to s_{max}, e.g. if sub-synchronous speed control is required down to a slip of 0·6, the effective stator-to-rotor turns ratio should also be 0·6.

The alternating voltage applied to the inverter terminals is sometimes reduced by means of a step-down transformer. This inverter transformer is essential with large a.c. supply voltages, since the slip-ring voltage is limited and may be too low for direct inversion into the a.c. supply. The transformer ratio is chosen in conjunction with the effective turns ratio of the motor so that the maximum controllable slip requires full inversion. This again maintains the power factor of the system at the highest possible value throughout the speed range.

Several circuit modifications have been devised in order to improve the power factor of the cascade drive. A special type of "through-pass" inverter has been developed to replace the phase-controlled inverter.(211) This inverter circuit can be phase-commutated or forced-commutated and returns slip power to the a.c. supply without drawing large amounts of reactive power. A reduction in reactive power consumption is also obtained by using a half-controlled inverter bridge of six thyristors in which one group of three thyristors is set for full inversion while the other group operates between full rectification and full inversion.(212) Motor-winding tap-changing has been suggested as a means of improving the power factor and reducing the converter rating in cascade drives with fan-type loads.(213) Similar advantages are achieved by using a changeover arrangement in which the slip-rings are switched onto a step-down transformer and an auxiliary diode bridge rectifier in the lower part of the speed range.(205) Capacitive compensation has also been suggested as a method of improving the power factor and giving increased torque at low speeds.(214)

10.3.2. *Operation with a Limited Sub-synchronous Speed Range*

In a sub-synchronous cascade drive which is controlled from standstill to full speed, the static converter must have the same kVA rating as the induction motor. When the sub-synchronous speed range is limited, however, the converter rating is reduced proportionately and the system power factor is improved. The induction motor is run-up from standstill in the usual manner with a starting resistor in the rotor circuit. At the lowest speed in the control range, the cascade converter is automatically switched-in by a speed detector, and the starting resistor is then disconnected.

The three-phase bridge rectifier has a voltage rating which is determined by the maximum working slip of the motor, since rotor e.m.f. and slip are proportional. The current rating of the rectifier bridge is determined by the maximum rotor current corresponding to the maximum output torque. With a constant-torque load, the rotor current, I_2, and the air-gap power, P_{ag}, are approximately constant. Maximum rotor voltage occurs at the greatest controllable slip, s_{max}, when, neglecting rotor losses, the cascade converter transfers the slip power, $s_{max}P_{ag}$, from the rotor to the supply. With a fan load, the torque varies as the square of the speed, and hence the rotor current is a maximum at full speed. The slip-ring voltage, however, is greatest at the lowest controllable speed, but the rectifier bridge must be rated for the maximum alternating voltage and current, even though they do not occur simultaneously.

The inverter bridge is rated for the a.c. supply voltage and the alternating current corresponding to the maximum output torque of the motor. In order to minimize the ratings of rectifier and inverter, the motor is designed with a transformation ratio such that the lowest controllable speed requires full inversion. This also optimizes the power factor of the system, as already explained. The inverter transformer which is required in the cascade circuit of high-voltage machines is, of course, only rated for the inverter kVA.

When the sub-synchronous speed range is limited, the power factor of the cascade drive is improved.[212] This is illustrated in Fig. 10.4, which is an approximate phasor diagram for a constant-torque loading of the motor. All harmonic currents are neglected, and the current phasors in the diagram represent the fundamental components of the non-sinusoidal

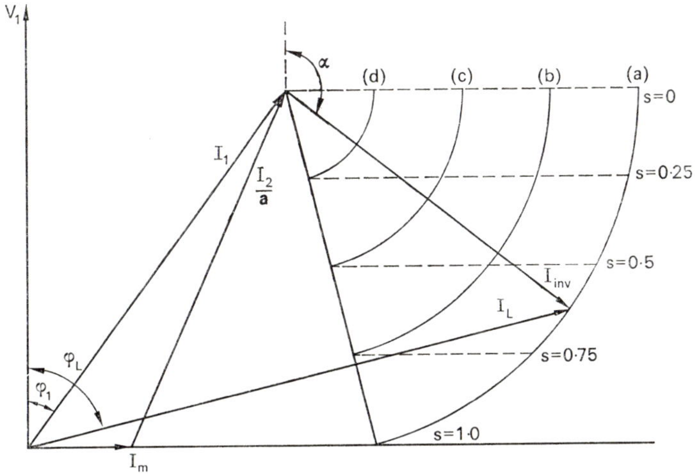

FIG. 10.4. Locus diagram of supply current for constant-torque operation and a speed control range: (a) from 0 to 100 per cent of rated speed; (b) from 25 to 100 per cent of rated speed; (c) from 50 to 100 per cent of rated speed; (d) from 75 to 100 per cent of rated speed.

waveforms. The fundamental stator current, $\mathbf{I}_1$, is obtained by combining the magnetizing component, $\mathbf{I}_m$, and the load component, $\mathbf{I}_2/a$, as in Fig. 10.4. The stator current lags the phase voltage, $\mathbf{V}_1$, by the phase angle, ϕ_1, and the tip of the $\mathbf{I}_1$ phasor lies inside the normal circle diagram of the motor with a short-circuited rotor. This is due to the commutating reactive power which is drawn by the diode bridge rectifier in the rotor circuit.[208–10] The fundamental component of inverter current, $\mathbf{I}_{inv}$, lags the supply voltage, $\mathbf{V}_1$, by the firing angle, α, if commutation overlap is neglected. The phasor sum of $\mathbf{I}_1$ and $\mathbf{I}_{inv}$ gives the total line current, $\mathbf{I}_L$, with the phase angle, ϕ_L.

In Fig. 10.4 a constant load torque is assumed and, consequently, the rotor current, $\mathbf{I}_2$, and the stator current, $\mathbf{I}_1$, do not vary appreciably. The inverter firing delay is assumed to be 165° at the lowest controllable speed, thus allowing 15° for reliable thyristor commutation. For speed control down to standstill, the current locus is given by curve (a). At standstill, the slip power returned to the supply is equal to the stator input power, if all losses are neglected. The total line current, $\mathbf{I}_L$, then lags the supply voltage by 90°. At a slip of 0·5, the slip power is half the stator power, and the active component of $\mathbf{I}_{inv}$ is half the active component of $\mathbf{I}_1$. In this way, the loci of Fig. 10.4 are obtained, showing the reduction in reactive power consumption as the speed range is limited.

A significant improvement in the system power factor and a reduction in the rating and cost of the cascade converter are achieved, therefore, when the sub-synchronous speed range is limited. The converter rating is then appreciably less than that of the static converter in an armature-controlled d.c. motor drive for the same horsepower. In the d.c. drive, the converter is rated for the maximum armature voltage and current, corresponding to full-load operation at maximum speed. In a static cascade drive with a speed control range from 70 per cent to 100 per cent of full speed, the converter kVA rating is only 30 per cent of the motor rating. The induction motor and static converter cascade system is, therefore, particularly suitable for drives requiring a limited speed range below synchronism. Typical applications are fan and pump drives, for which the power output varies as the cube of the speed and hence a limited sub-synchronous speed range is adequate.

10.3.3. *The Influence of Harmonics*

In the cascade converter, the slip power is rectified to d.c., filtered, and inverted to a.c. The rectification process produces non-sinusoidal rotor currents which induce associated harmonic currents in the stator winding. These current harmonics cause additional harmonic torques and losses which necessitate some de-rating of the induction motor. An asynchronous harmonic torque is produced by the interaction between a time harmonic rotor flux and the corresponding induced stator current. The asynchronous torques are usually small in magnitude.[215] Space harmonic fluxes also produce synchronous locking torques at certain speeds, but their effect is negligible in a good machine design.

The smoothing reactor in the d.c. link is necessary for satisfactory operation of the inverter, and a large inductance is required to limit the flow of harmonic current in the rotor winding. The harmonic currents in stator and rotor cause harmonic copper losses which slightly reduce the motor efficiency. In order to limit these harmonic losses to a given percentage of the machine losses, the value of the filter inductance should be increased as the rated d.c. link current of the system decreases.[215] Converter losses must also be taken into consideration when calculating the overall system efficiency. In practice, the cascade converter losses with a mercury-arc inverter have been quoted as[208]

$$P_{\text{conv}} \simeq (0{\cdot}046 \text{ to } 0{\cdot}08)\, s_{\max} P_{ag}$$

where the lower limit applies to a 2-MW static converter, and the upper limit is for a 100-kW converter. In a static converter cascade system, the converter losses are considerably lower than in a conventional Scherbius drive which has three rotating machines in the cascade circuit. The full-load efficiency near maximum speed in a static converter cascade system is only a few per cent less than the normal motor efficiency on full-load with a short-circuited rotor.

The harmonic losses increase the temperature of the induction motor and reduce its rating. The commutating reactive power, which is drawn through the motor by the bridge rectifier, also reduces the useful power rating of the motor. As a result, the induction motor must be de-rated by 10 per cent, or more, when it is employed in a static cascade connection.

10.3.4. *Closed-loop Control*

The cascade drive, like the separately excited d.c. motor, has a linear torque–current relationship, and speed control is normally obtained by means of a double-loop control system which is similar to that used in single-converter d.c. motor drives. Figure 10.5 shows a block diagram of the control scheme. The inner control loop adjusts the current in the cascade circuit by variation of the thyristor firing angle, and this current determines the motor torque. The actual cascade current is measured by current transformers on the a.c. side of the inverter, and the desired current value is set by the outer speed loop, which measures the

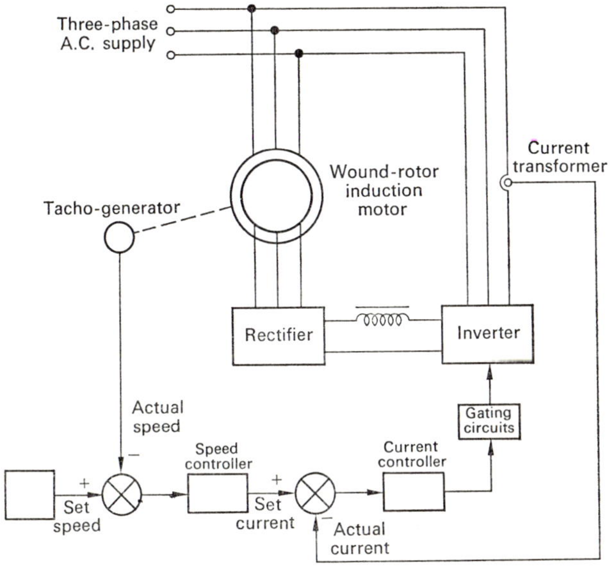

FIG. 10.5. Block diagram of control system for closed-loop operation.

difference between the desired speed, as set on a potentiometer, and the actual tacho-generator value. Thus, a speed error produces a motor torque which reduces the error. Current limitation is readily incorporated by limiting the reference current value delivered by the speed-error amplifier. In this manner, the cascade current may be limited to 120 per cent, say, of rated current, even under stalled conditions. If the desired speed greatly exceeds the measured value, the motor accelerates rapidly with maximum current and torque until the desired speed is approached. The current and torque are then reduced automatically to the values required by the load. For a sudden reduction in demanded speed, the current is reduced to a low level, and the braking action of the load reduces the motor speed.

In the cascade drive, the control system is obviously much simpler than in other variable-speed induction motor drives in which the rotor slip is measured and controlled. The double-loop control system is readily stabilized, and the speed regulation on load may be reduced by means of a high-gain speed-error amplifier. In this manner, a speed droop from no-load to full-load of 0·1 per cent, or less, may be achieved.

10.4. SUPER-SYNCHRONOUS STATIC CONVERTER CASCADE

The static converter circuit of Fig. 10.2 achieves sub-synchronous speed control by the removal of slip power from the rotor circuit. In fact, super-synchronous speeds may also be obtained by reversing the direction of power flow in the cascade circuit, thereby feeding additional power into the rotor. This is readily achieved using rotating machinery, since all electrical machines are inherently capable of motor or generator operation. In order to reverse the power flow in the converter circuit of Fig. 10.2, the rectifier must operate as an inverter and vice versa. For speed control on either side of synchronism, the converter cascade therefore needs two phase-controlled thyristor bridges, one operating at slip frequency as a rectifier or inverter, while the other operates at network frequency as an inverter or rectifier. The replacement of six diodes by six thyristors increases the converter cost appreciably, and also necessitates the introduction of a slip-frequency gating circuit. Difficulty is experienced near synchronism when the slip-frequency e.m.f.s are insufficient for phase commutation, and special connections or forced-commutation methods are necessary for the passage through synchronism.[216–18] Thus, the provision of super-synchronous speed control unduly complicates the static converter cascade system and nullifies the advantages of simplicity and economy which are inherent in a purely sub-synchronous drive.

10.5. STATIC CONTROL OF ROTOR RESISTANCE

Intermittent low-speed operation of a wound-rotor induction motor can be simply obtained by the introduction of external rotor resistance to dissipate the slip power. Speed control is obtained by mechanical variation of the external resistance, and the drive is suitable for crane and hoist applications where sustained low-speed running is not required. Nowadays, a high-frequency thyristor chopper circuit allows the external rotor resistance to be varied statically and steplessly, and provides a low-cost variable-speed drive with a good dynamic response.

Figure 10.6 shows a suitable circuit in which the rotor slip power is rectified in a diode bridge rectifier and fed through a smoothing reactor to a resistor, R. A single thyristor in

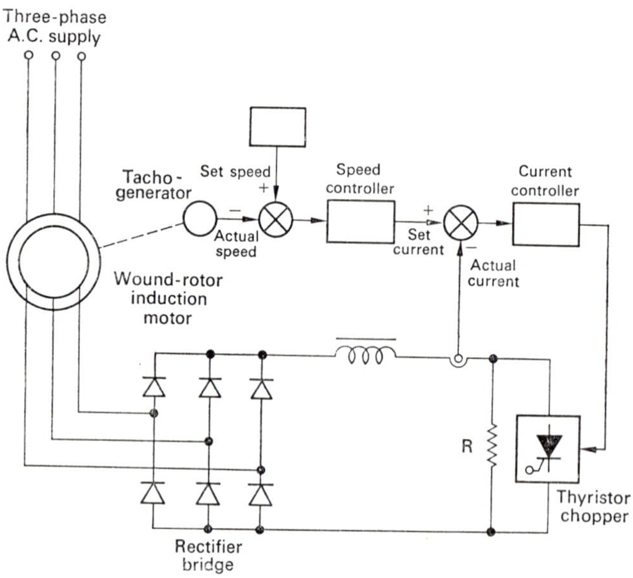

FIG. 10.6. Induction motor speed control by static variation of external rotor resistance.

parallel with the resistor is switched on and off, at a frequency of about 1 kHz, by a chopper circuit of the type described in Section 3.6.2. The ratio of on-time to off-time determines the effective value of rotor resistance and thus controls the motor speed by altering its torque-speed characteristic. By introducing a capacitor in series with the external resistor, it is possible to obtain a variation in the effective resistance from zero to infinity, thus permitting a wider range of speed control.[219–20] This type of rotor circuit may also be used in combination with the stator-voltage-control circuits of Chapter 9.[221] When the load torque is small, speed control is obtained by variation of the stator voltage, and rotor resistance control is used in the high-torque range.

APPENDIX A

Basic Theory of the Polyphase Induction Motor

THE induction motor has a symmetrical polyphase stator winding which is connected to a balanced polyphase a.c. supply. The wound-rotor machine also has a polyphase winding on the rotor, and the terminals are connected to slip-rings which are usually short-circuited. In a squirrel-cage motor, the individual rotor conductors are short-circuited by low-resistance end-rings. Thus, the rotor winding of the induction motor has no external connection, and receives its supply by induction from the stator. The stator currents establish a rotating air-gap m.m.f. wave, and this produces a flux wave of constant amplitude which rotates at uniform speed, moving through a double pole pitch in each cycle of the supply frequency. The speed of rotation of the field is therefore $\omega_1 = 2\pi f_1/p$ rad/sec, or $n_1 = 60f_1/p$ rev/min, where p is the number of stator pole-pairs and f_1 is the supply frequency in cycles/sec or Hz. This is the synchronous speed of the induction motor.

For normal operation, the supply voltage and frequency are both constant. When the machine is at rest, the synchronously rotating field induces e.m.f.s of supply frequency in the stationary rotor. These rotor e.m.f.s are indicated in Fig. A.1 (a) for a two-pole cage-rotor machine with an air-gap flux rotating anti-clockwise. The rotor conductors are short-cir-

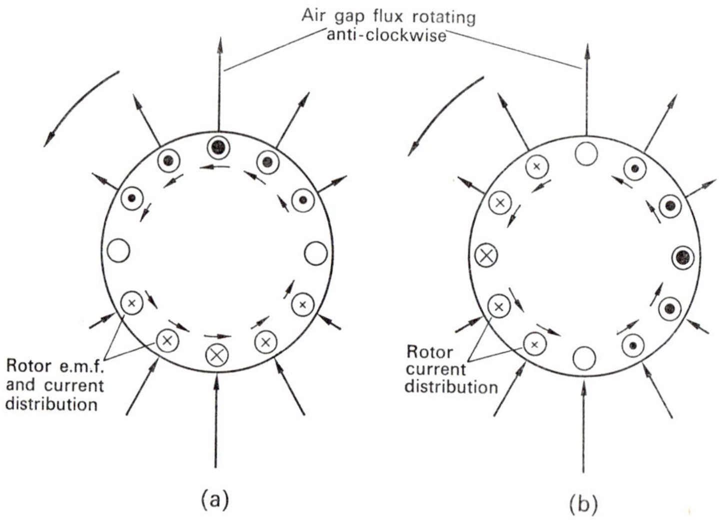

FIG. A.1. Torque production in a squirrel-cage rotor: (a) maximum rotor torque in a purely resistive rotor circuit: (b) zero rotor torque in a purely inductive rotor circuit.

cuited at each end and, if the circuit is purely resistive, the current and e.m.f. distributions are identical. The torque contribution of each rotor conductor is proportional to the product of the conductor current and the local flux density. In Fig. A.1 (a) the rotor conductor with the largest current lies in the region of greatest flux density, and each conductor produces a torque contribution in the anti-clockwise direction as indicated by the arrows. These are the optimum conditions for the production of electromagnetic torque. If the rotor circuit is purely inductive, the resultant torque is zero, as indicated in Fig. A.1 (b). In practice, the rotor power factor is always greater than zero, and hence a motor torque is developed which causes the machine to run-up from standstill in the same direction as the air-gap field.

Assume the motor is running at a speed n which is less than the synchronous speed n_1. The difference, $(n_1 - n)$, is called the slip speed of the machine and is usually expressed as a fraction of the synchronous speed to give the fractional slip, $s = (n_1 - n)/n_1$. The air-gap field now sweeps past the rotor at a relative speed of $(n_1 - n)$. The rotor frequency is therefore

$$f_2 = \left(\frac{n_1 - n}{n_1}\right) f_1 = sf_1. \tag{A.1}$$

For normal loading conditions it can be assumed, to a good approximation, that the air-gap flux is constant. The magnitude of the rotor e.m.f. is then proportional to the fractional slip s. If the rotor runs in exact synchronism with the air-gap field, the induced e.m.f. is zero and the motor develops zero torque. In practice, even on no-load, a small motor torque is required to overcome windage and friction, and the motor runs at a speed which is slightly less than the synchronous speed. When load is applied to the motor shaft, the speed drops further below synchronism and larger rotor currents are induced. This allows the development of the increased driving torque demanded by the load. When a rotor current flows, an additional component of stator current is induced. As in a transformer, this load component of stator current neutralizes the rotor m.m.f. and leaves the resultant air-gap flux practically unaltered, as assumed above.

Torque-speed Characteristic

Assuming constant air-gap flux, the rotor e.m.f. is proportional to the fractional slip and, in a purely resistive rotor circuit, the current and torque are also proportional to slip. Thus, the induction motor has a shunt torque-speed characteristic and the speed decreases linearly with increasing load torque. This analysis is valid for low-slip operation near synchronous speed when the rotor frequency is only a few Hz. At these low frequencies, the rotor leakage reactance is negligible, but it becomes more significant as the speed falls and the slip frequency f_2 increases. The rotor current is reduced as a result of the increased rotor impedance and, in addition, the current lags the e.m.f. by the phase angle ϕ_2 of the rotor circuit. In general, the rotor current can be resolved into two components, one component in phase with the rotor e.m.f. and the other component lagging by 90°. The in-phase current has the rotor distribution of Fig. A.1 (a) and the quadrature current has the distribution of Fig. A.1 (b). The latter distribution produces no resultant torque, and hence the distribution of Fig. A.1 (a) represents the torque-producing component of rotor current. Thus, if I_2 is the r.m.s. current in a rotor conductor and ϕ_2 is the rotor phase angle, the electromagnetic torque is determined by the in-phase rotor current $I_2 \cos \phi_2$. The rotor torque T is also proportional to the air-gap flux per pole Φ, and hence

$$T = K\Phi I_2 \cos \phi_2 \tag{A.2}$$

where K is a constant. This analysis is also valid for a wound-rotor machine if I_2 is the r.m.s. phase current in the rotor.

Equivalent Circuit

The equivalent circuit of the induction motor is very similar to the usual transformer equivalent circuit, since the induction motor is essentially a transformer with a rotating secondary winding. As in a static transformer, the primary or stator current establishes a mutual flux which links the secondary or rotor winding, and also a leakage flux which links only the primary winding. This leakage flux induces a primary e.m.f. which is proportional to the rate of change of primary current, and its effect may be represented, in the usual manner, by a series leakage reactance X_1 in each stator phase (Fig. A.2). R_1 is the stator

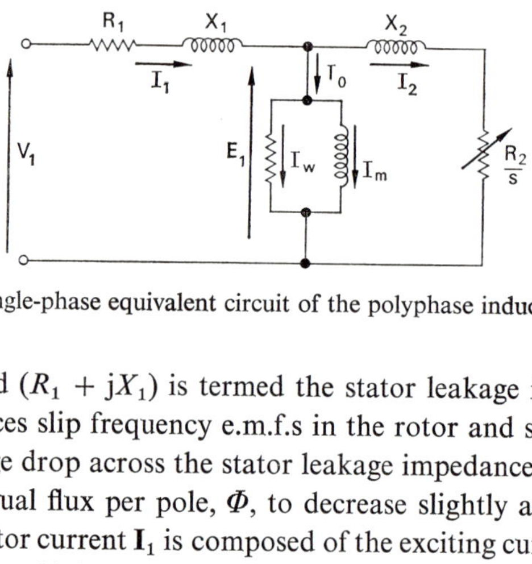

FIG. A.2. Single-phase equivalent circuit of the polyphase induction motor.

resistance per phase and $(R_1 + jX_1)$ is termed the stator leakage impedance. The mutual flux in the air gap induces slip frequency e.m.f.s in the rotor and supply frequency e.m.f.s in the stator. The voltage drop across the stator leakage impedance causes the stator e.m.f. per phase, E_1, and mutual flux per pole, Φ, to decrease slightly as load is applied to the motor. The resultant stator current $\mathbf{I}_1$ is composed of the exciting current, and the load component of stator current which cancels the m.m.f. due to the rotor current. The exciting current $\mathbf{I}_o$ consists of the magnetizing and core-loss components, $\mathbf{I}_m$ and $\mathbf{I}_w$ respectively.

In deriving the rotor equivalent circuit, the actual phase-wound or squirrel-cage rotor winding is considered to be replaced by an equivalent short-circuited rotor winding having the same number of turns and the same winding arrangement as the stator. This is equivalent to the usual transformer procedure of referring secondary quantities to the primary. At standstill, the induced e.m.f. per phase in the equivalent rotor is equal to the stator e.m.f., E_1, and the rotor frequency equals the supply frequency, f_1. When the motor runs with a slip s, the rotor e.m.f., $E_2 = sE_1$, and the rotor frequency $f_2 = sf_1$. If R_2 is the equivalent rotor resistance per phase, and X_2 is the rotor leakage reactance per phase at standstill, then the rotor current is given by

$$\mathbf{I}_2 = \frac{\mathbf{E}_2}{R_2 + \mathrm{j}sX_2} = \frac{s\mathbf{E}_1}{R_2 + \mathrm{j}sX_2} \tag{A.3}$$

and hence

$$\mathbf{I}_2 = \frac{\mathbf{E}_1}{(R_2/s) + \mathrm{j}X_2}. \tag{A.4}$$

In equation (A.3), all rotor quantities are at slip frequency, but in equation (A.4) they are at supply frequency. This shows that the rotor current I_2 is unaltered in magnitude if the rotor is brought to standstill and the resistance increased from R_2 to R_2/s. The rotor equivalent circuit may therefore be joined directly to the stator circuit, as in Fig. A.2, to give the complete equivalent circuit for one phase of the motor.

Torque Equation

At a slip s, the rotor power loss in the equivalent circuit is $I_2^2R_2/s$ watts per phase, whereas in the actual machine the rotor copper loss is $I_2^2R_2$ watts per phase. The additional power loss in the equivalent circuit is the electrical equivalent of the mechanical power output of the motor. If P_{mech} denotes the gross mechanical power output including windage and friction losses, then

$$P_{\mathrm{mech}} = m_1\,[(I_2^2R_2/s) - (I_2^2R_2)] = m_1I_2^2R_2\left(\frac{1-s}{s}\right)$$

where m_1 is the number of stator phases.

If ω is the mechanical angular velocity of the rotor and T is the electromagnetic torque

$$T\omega = m_1I_2^2R_2\left(\frac{1-s}{s}\right)$$

and

$$T = \frac{m_1I_2^2R_2}{\omega}\left(\frac{1-s}{s}\right).$$

This is the internal motor torque which is greater than the useful shaft torque by the amount required to overcome the windage and friction torques.

Since the synchronous angular velocity is given by $\omega_1 = \omega/(1-s) = 2\pi f_1/p$, the torque equation can be rewritten as

$$T = \frac{m_1I_2^2R_2}{s\omega_1} \tag{A.5}$$

or

$$T = \frac{pm_1}{2\pi f_1}\,(I_2)^2\,\frac{R_2}{s}. \tag{A.6}$$

Power Division in the Rotor

It can be seen from the equivalent circuit that the total electrical power input to the rotor across the air gap from the stator is

$$P_{ag} = m_1 I_2^2 R_2/s. \tag{A.7}$$

This is divided between the mechanical power output, P_{mech}, and the rotor copper loss, P_2. Thus,

$$P_{ag} = P_{mech} + P_2$$

where

$$P_{mech} = T\omega$$

and

$$P_2 = m_1 I_2^2 R_2.$$

Combining equations (A.5) and (A.7) gives

$$P_{ag} = T\omega_1 \tag{A.8}$$

and hence the total electrical power input to the rotor is equal to the internal mechanical torque multiplied by the synchronous angular velocity.

APPENDIX B

The Two-transistor Analogue of the Thyristor

THE behaviour of a thyristor is best understood by considering that the *p–n–p–n* structure is formed by two complementary transistors, a *p–n–p* and an *n–p–n*, having a common collector junction. As shown in Fig. B.1, the collector of each transistor is cross-coupled to the base of the other and, consequently, there is a positive feedback action which allows each

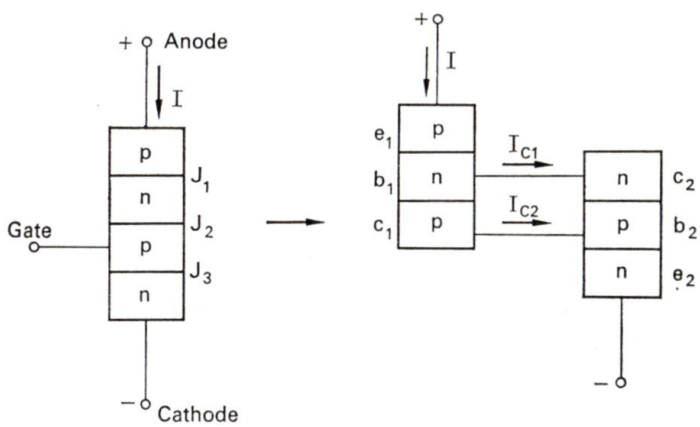

FIG. B.1. The two-transistor analogue of the thyristor.

transistor to drive the other into saturation. When the thyristor has a positive anode-to-cathode voltage, each transistor is biased in the conventional manner and has a common base current gain, α, which expresses the ratio of the collector and emitter currents. Thus, for the *p–n–p* transistor, α_1 is the fraction of the hole current injected at the emitter e_1 that reaches the collector c_1. If I is the external current entering emitter e_1, then the collector current is given by

$$I_{c1} = \alpha_1 I + I_{1co}$$

where I_{1co} is the collector leakage current of the transistor.

Similarly, the *n–p–n* transistor has a current gain α_2, which is the fraction of electron current injected at the emitter e_2 that reaches collector c_2.

Thus $I_{c2} = \alpha_2 I + I_{2co}$ where I_{2co} is the leakage current of the *n–p–n* transistor.

The sum of the two collector currents is equal to the current flowing across the common collector junction J_2 of the thyristor. This current obviously must be equal to the external circuit current I:

$$\therefore I = I_{c1} + I_{c2} = (\alpha_1 + \alpha_2) I + I_{co}$$

where I_{co} is the total leakage current of the thyristor.

Solving for I gives

$$I = \frac{I_{co}}{1 - (\alpha_1 + \alpha_2)}. \qquad \text{(B.1)}$$

If the current gains α_1 and α_2 are small, the thyristor current is only slightly greater than the leakage current I_{co}, and hence the device is in its forward-blocking state. The thyristor remains in this "off" condition unless $(\alpha_1 + \alpha_2)$ is increased by one of the methods about to be described. When $(\alpha_1 + \alpha_2)$ approaches unity, regenerative action takes place and the thyristor current tends to infinity. The resistance of the external circuit must limit the flow of current, since both transistors are saturated and all junctions are forward-biased. Under these conditions, the *p–n–p–n* device has a very low impedance and is in the forward-conducting or "on" state.

Gate Triggering

The current gain, α, of a transistor is a function of the emitter current (Fig. B.2) and, consequently, if the base-emitter junction of either transistor is forward-biased, the sum of the current gains $(\alpha_1 + \alpha_2)$ will increase, causing the thyristor to "turn on". Usually the base of the *n–p–n* transistor is used as the gate terminal into which current is injected in

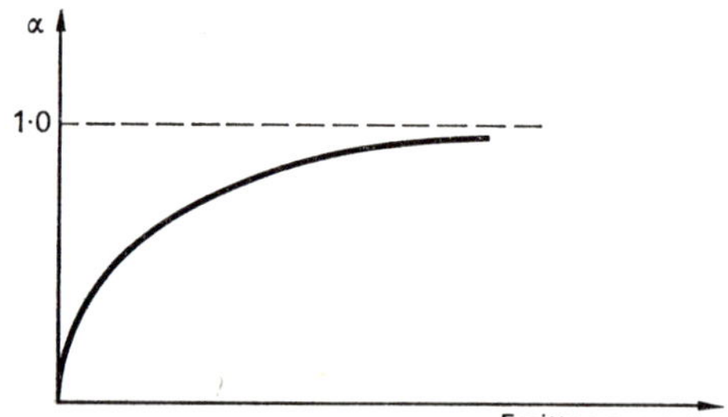

FIG. B.2. Variation of transistor current gain with emitter current.

order to switch the thyristor into the conducting state. The thyristor continues to conduct when the gate current is removed, provided the load current exceeds a certain low value called the holding or sustaining current. When the load current is less than the holding current, $(\alpha_1 + \alpha_2) < 1$, and the thyristor reverts to the forward blocking mode.

Forward Breakover

An avalanche breakdown occurs in a transistor when the collector-to-emitter voltage is increased excessively. This is due to the leakage current carriers which arrive at the collector junction with sufficient energy to dislodge additional carriers. The avalanche multiplication causes a sharp increase in transistor current, and hence the current gain, α, also increases.

The forward blocking voltage of a thyristor can be gradually increased until avalanche breakdown occurs at the common collector junction J_2. The sum of the current gains then reaches unity and the thyristor switches on. The forward blocking voltage just prior to switching is known as the forward breakover voltage and is denoted by $V_{(BO)}$. In practice, this method of firing a thyristor is rarely used, and gate triggering is the mechanism which is normally employed.

DV/DT Turn On

A high rate of increase of forward anode-to-cathode voltage (dv/dt) can turn on a thyristor, even though there is no gate signal and the forward breakover voltage is not exceeded. When the thyristor is in the forward-blocking state, the applied voltage appears as a reverse bias on junction J_2, since junctions J_1 and J_3 are forward-biased. The reverse-biased *p–n* junction is equivalent to a capacitor, since the mobile current carriers are removed from the vicinity of J_2, leaving exposed charges on either side of the junction. When the applied voltage is increased, there is a further movement of mobile charges, and this is equivalent to the charging current of the capacitor. If the forward blocking voltage has a high rate of increase, there is a significant charge movement away from the central junction, J_2. Additional carriers are injected from the end-regions to neutralize the excess charge. These transient currents will trigger the thyristor into conduction if $(\alpha_1 + \alpha_2)$ approaches unity. In practice, thyristors are available which can withstand dv/dt values of several hundred volts per microsecond.

References

1. ABRAHAM, L. ,FÖRSTER, J. and SCHLIEPHAKE, G., A.C. motor supply with thyristor converters, *I.E.E.E. Trans. on Ind. and Gen. Applic.* **IGA-2**, 334 (1966).
2. KUSKO, A., What's available in adjustable speed a.c. drives, *Control Engng.* **15**, 58 (1968).
3. DINGER, E.H., Digital applications of a.c. motor controls, *I.E.E.E. Trans. on Ind. and Gen. Applic.* **IGA-5**, 291 (1969).
4. HELMICK, C.G., Present status of adjustable-frequency drive applications, *I.E.E.E. Ind. and Gen. Applic. Group Second Annual Meeting* 275 (1967).
5. MARTIN, A.M., Thyristors for motor control—changing economics, *Electl Times* **157**, 63 (1970).
6. KANNGIESSER, K.-W., Umrichter zur Speisung von Drehfeldmaschinen, *Elektrotech. Z. Ausg. A* **85**, 673 (1964).
7. SCHNÖRR, R., Der Drehstrommotor mit Umrichterspeisung, *V.D.E. Fachber.* **23**, 225 (1964).
8. SCHÖNUNG, A., Varying the speed of three-phase motors by means of static frequency changers, *Brown Boveri Rev.* **51**, 540 (1964).
9. HUMPHREY, A.J., Precise speed control with inverters, *I.E.E.E. Industrial Static Power Conversion Conf.* 78 (1965).
10. HEUMANN, K., Variable frequency speed control of induction motors, Paper 4D, *I.F.A.C. Congress, London* (1966).
11. STEMMLER, H., The use of static converters to vary the speed of three-phase drives without losses, *Brown Boveri Rev.* **54**, 217 (1967).
12. DAVIS, E.W. and MEIER, R.W., Application of inverter drives, *I.E.E.E. Trans. on Ind. and Gen. Applic.* **IGA-5**, 45 (1969).
13. VERES, R.P., New inverter supplies for high horsepower drives, *I.E.E.E. Trans. on Ind. and Gen. Applic.* **IGA-6**, 121 (1970).
14. ABRAHAM, L. and KOPPELMANN, F., Die Zwangskommutierung, ein neuer Zweig der Stromrichtertechnik, *Elektrotech. Z. Ausg. A* **87**, 649 (1966).
15. HEUMANN, K., Elektrotechnische Grundlagen der Zwangskommutierung—Neue Möglichkeiten der Stromrichtertechnik, *Elektrotechnik MaschBau* **84**, 99 (1967).
16. FLAIRTY, C.W., A 50-kVA adjustable frequency 24 phase controlled rectifier inverter, *Direct Curr.* **6**, 278 (1961).
17. USHER, T.E. and BECK, C.D., Adjustable speed with a.c. motors and adjustable frequency inverters, *A.I.E.E. Conf.* Paper No. DP 62-981 (1962).
18. BYSTRON, K. and MEYER, M., Kontaktlose, drehzahlregelbare Umrichtermaschinen für hohe Drehzahlen, *Siemens-Z.* **37**, 660 (1963).
19. ABRAHAM, L., HEUMANN, K. and KOPPELMANN, F., Wechselrichter zur Drehzahlsteuerung von Käfigläufermotoren, *A.E.G. Mitt.* **54**, 89 (1964).
20. BRADLEY, D.A., CLARKE, C.D., DAVIS, R.M. and JONES, D.A., Adjustable-frequency inverters and their application to variable-speed drives, *Proc. Instn. elect. Engrs* **111**, 1833 (1964).
21. DÜLL, E.H. and GOLDE, E., Dreiphasenumrichter mit Gleichstrom-Zwischenkreis 50/50 bis 500 Hz, *A.E.G. Mitt.* **54**, 165 (1964).
22. HULTSTRAND, K.A., Static adjustable frequency drives, *Control Engng* **11**, 57 (1964).
23. MACDONALD, I.M., A static inverter, wide range, adjustable speed drive, *I.E.E.E. Int. Conv. Rec.* Pt. 4, **12**, 34 (1964).
24. SCHÖNUNG, A. and STEMMLER, H., Static frequency changers with "subharmonic" control in conjunction with reversible variable-speed a.c. drives, *Brown Boveri Rev.* **51**, 555 (1964).
25. WARD, E.E., Inverter suitable for operation over a range of frequency, *Proc. Instn elect. Engrs* **111**, 1423 (1964).
26. BYSTRON, K. and MEISSEN, W., Drehzahlsteuerung von Drehstrommotoren über Zwischenkreisumrichter, *Siemens-Z.* **39**, 254 (1965).

27. KING, K. G., Variable frequency thyristor inverters for induction motor speed control, *Direct Curr.* **10**, 26 (1965).
28. OHNO, E. and AKAMATSU, M., Variable frequency s.c.r. inverter with an auxiliary commutation circuit, *I.E.E.E. Trans. on Magnetics* **MAG-2**, 25 (1966).
29. TURNBULL, F. G., Wide range, impulse commutated, static inverter with a fixed commutation circuit, *I.E.E.E. Ind. and Gen. Applic. Group First Annual Meeting* 475 (1966).
30. MOKRYTZKI, B., Pulse width modulated inverters for a.c. motor drives, *I.E.E.E. Trans. on Ind. and Gen. Applic.* **IGA-3**, 493 (1967).
31. LI, K. Y. G., New 3-phase inverter circuit, *Proc. Instn elect. Engrs* **115**, 1677 (1968).
32. ESPELAGE, P. M., CHIERA, J. A. and TURNBULL, F. G., A wide range, static inverter suitable for a.c. induction motor drives, *I.E.E.E. Trans. on Ind. and Gen. Applic.* **IGA-5**, 438 (1969).
33. HEUMANN, K., Development of inverters with forced commutation for a.c. motor speed control up to the megawatt range, *I.E.E.E. Trans. on Ind. and Gen. Applic.* **IGA-5**, 61 (1969).
34. JOHNSTON, R. W., Modulating inverter system for variable-speed induction motor drive, *I.E.E.E. Trans. on Power App. and Systems* **PAS-88**, 81 (1969).
35. RISBERG, R. L., A wide speed range inverter fed induction motor drive, *I.E.E.E. Ind. and Gen. Applic. Group Fourth Annual Meeting* 629 (1969).
36. ROBERTSON, S. D. T. and HEBBAR, K. M., A tuned circuit commutated inverter, Instn elect. Engrs Conf. Publ. No. 53, *Power Thyristors and Their Applications* 297 (1969).
37. WILLIAMSON, K. H., Simple 3 phase a.c. motor control system for motors below 5 h.p., Instn elect. Engrs Conf. Publ. No. 53, *Power Thyristors and Their Applications* 320 (1969).
38. CHANDLER, E. F. and PETERS, F. N., The wide speed range inverter, *I.E.E.E. Trans. on Ind. and Gen. Applic.* **IGA-6**, 19 (1970).
39. GRIFFITH, D. C. and ULMER, R. M., A semiconductor variable-speed a.c. motor drive, *Electl Engng, New York* **80**, 350 (1961).
40. HECK, R. and MEYER, M., A static-frequency-changer-fed squirrel-cage motor drive for variable speed and reversing, *Siemens Rev.* **30**, 401 (1963).
41. ANNIÉS, B., Steuerumrichter für Käfigläufermotoren, *A.E.G. Mitt.* **54**, 123 (1964).
42. GUYESKA, J. C. and JORDAN, H. E., Static a.c. variable frequency drive, I.E.E.E. Conf. Paper No. CP 64-391 (1964).
43. BOWLER, P., The application of a cycloconverter to the control of induction motors, Instn elect. Engrs Conf. Publ. No. 17, *Power Applications of Controllable Semiconductor Devices* 137 (1965).
44. SLABIAK, W. and LAWSON, L. J., Precise control of a three-phase squirrel-cage induction motor using a practical cycloconverter, *I.E.E.E. Trans. on Ind. and Gen. Applic.* **IGA-2**, 274 (1966).
45. HAMILTON, R. A. and LEZAN, G. R., Thyristor adjustable frequency power supplies for hot strip mill run-out tables, *I.E.E.E. Trans. on Ind. and Gen. Applic.* **IGA-3**, 168 (1967).
46. ALGER, P. L., *The Nature of Induction Machines*, Gordon & Breach, New York, 1965.
47. LAWRENSON, P. J. and AGU, L. A., Theory and performance of polyphase reluctance machines, *Proc. Instn elect. Engrs* **111**, 1435 (1964).
48. CRUICKSHANK, A. J. O., MENZIES, R. W. and ANDERSON, A. F., Axially laminated anisotropic rotors for reluctance motors, *Proc. Instn elect. Engrs* **113**, 2058 (1966).
49. CHALMERS, B. J. and MULKI, A. S., New reluctance motors with unlaminated rotors, *Proc. Instn elect. Engrs* **117**, 2271 (1970).
50. FONG, W. and HTSUI, J. S. C., New type of reluctance motor, *Proc. Instn elect. Engrs* **117**, 545 (1970).
51. LAWRENSON, P. J., Development and application of reluctance motors, *Electronics and Power* **11**, 195 (1965).
52. SCOTT, A., Reluctance motor reborn, *Engineering, London* **208**, 626 (1969).
53. LAWRENSON, P. J. and GUPTA, S. K., Developments in the performance and theory of segmental-rotor reluctance motors, *Proc. Instn elect. Engrs* **114**, 645 (1967).
54. GUTZWILLER, F. W. (Ed.), *Silicon Controlled Rectifier Manual*, 4th ed., General Electric Co., New York, 1967.
55. GENTRY, F. E., GUTZWILLER, F. W., HOLONYAK, N. and VON ZASTROW, E. E., *Semiconductor Controlled Rectifiers*, Prentice-Hall, New Jersey, 1964.
56. HEUMANN, K. and STUMPE, A. C., *Thyristoren*, Teubner, Stuttgart, 1969.
57. GRIFFIN, A. and RAMSHAW, R. S., *The Thyristor and its Applications*, Chapman & Hall, London, 1965.
58. MAPHAM, N. W., The classification of s.c.r. inverter circuits, *I.E.E.E. Int. Conv. Rec.* Pt. 4, **12**, 99 (1964).
59. BEDFORD, B. D. and HOFT, R. G., *Principles of Inverter Circuits*, J. Wiley, New York, 1964.
60. SHEPHERD, W., Steady-state analysis of the resistance-inductance circuit controlled by silicon controlled rectifiers, *I.E.E.E. Trans. on Ind. and Gen. Applic.* **IGA-1**, 259 (1965).

61. ADEM, A.A., Speed controls for universal motors, General Electric Co., New York, Application Note 200.47 (1966).
62. RISSIK, H., *The Fundamental Theory of Arc Converters*, Chapman & Hall, London, 1939.
63. SCHAEFER, J., *Rectifier Circuits: Theory and Design*, J.Wiley, New York, 1965.
64. KHASAYEV, O.I., Operation of an induction motor from a transistor frequency converter, *Elect. Technol. USSR* **3**, 441 (1961).
65. ANDERSON, D.L., WILLIS, A.E. and WINKLER, C.E., Advanced static inverter utilizing digital techniques and harmonic cancellation, N.A.S.A. Technical Note D-602 (1962).
66. COREY, P.D., Methods for optimizing the waveform of stepped-wave static inverters, A.I.E.E. Conf. Paper No. CP 62-1147 (1962).
67. KERNICK, A., ROOF, J.L. and HEINRICH, T.M., Static inverter with neutralization of harmonics, *Trans. A.I.E.E.*, Pt. 2 *(Applic. and Ind.)* **81**, 59 (1962).
68. TURNBULL, F.G., Selected harmonic reduction in static d.c.–a.c. inverters, *I.E.E.E. Trans. on Communication and Electronics* **83**, 374 (1964).
69. WAGNER, C.F., Parallel inverter with resistive load, *Trans. A.I.E.E.* **54**, 1227 (1935).
70. WAGNER, C.F., Parallel inverter with inductive load, *Trans. A.I.E.E.* **55**, 970 (1936).
71. McMURRAY, W. and SHATTUCK, D.P., A silicon-controlled-rectifier inverter with improved commutation, *Trans. A.I.E.E.*, Pt. 1 *(Communication and Electronics)* **80**, 531 (1961).
72. McMURRAY, W., SCR inverter commutated by an auxiliary impulse, *I.E.E.E. Trans. on Communication and Electronics* **83**, 824 (1964).
73. DEWAN, S.B. and DUFF, D.L., Optimum design of an input-commutated inverter for a.c. motor control, *I.E.E.E. Trans. on Ind. and Gen. Applic.* **IGA-5**, 699 (1969).
74. DEWAN, S.B. and DUFF, D.L., Practical considerations in the design of commutation circuits for choppers and inverters, *I.E.E.E. Ind. and Gen. Applic. Group Fourth Annual Meeting* 469 (1969).
75. SATO, N., Characteristics of s.c.r. three-phase improved bridge inverter, *Electl Engng in Japan* **84**, 30 (1964).
76. MORGAN, R.E., Basic magnetic functions in converters and inverters including new soft commutation, *I.E.E.E. Trans. on Ind. and Gen. Applic.* **IGA-2**, 58 (1966).
77. HUMPHREY, A.J., Inverter commutation circuits, *I.E.E.E. Trans. on Ind. and Gen. Applic.* **IGA-4**, 104 (1968).
78. HEUMANN, K., Pulse control of d.c. and a.c. motors by silicon controlled rectifiers, *I.E.E.E. Trans. on Communication and Electronics* **83**, 390 (1964).
79. LANDIS, J.P., Static Inverter A.C. Motor Drives, The Louis Allis Co., U.S.A.
80. DEWAN, S.B. and ROSENBERG, S.A., Output voltage in three-phase pulsewidth-modulated inverters, *I.E.E.E. Trans. on Ind. and Gen. Applic.* **IGA-6**, 570 (1970).
81. ABRAHAM, L., HEUMANN, K., KOPPELMANN, F. and PATZSCHKE, U., Electronic power pulsing of electric motor drives, *A.E.G. Prog.* 149 (1965).
82. FISHER, N.A., Application of re-rated induction motors and pulse-width modulated inverters to adjustable speed drive requirements, *I.E.E.E. Ind. and Gen. Applic. Group Second Annual Meeting* 473 (1966) (abstract only).
83. TURNBULL, F.G., A carrier frequency gating circuit for static inverters, converters, and cycloconverters, *I.E.E.E. Trans. on Magnetics* **MAG-2**, 14 (1966).
84. SCHENKEL, M., Eine unmittelbare asynchrone Umrichtung für niederfrequente Bahnnetze, *Elekt. Bahn.* **8**, 69 (1932).
85. BOSCH, M. and KASPEROWSKI, O., Die Steuerumrichter-Versuchsanlage der Siemens-Schuckertwerke A.-G. im Reichsbahn-Saalachkraftwerk, *Elekt. Bahn.* **11**, 235 (1935).
86. ALEXANDERSON, E.F.W. and MITTAG, A.H., The "thyratron" motor, *Electl Engng, New York* **53**, 1517 (1934).
87. BEILER, A.H., The thyratron motor at the Logan plant, *Electl Engng, New York* **57**, 19 (1937).
88. CHIRGWIN, K.M., STRATTON, L.J. and TOTH, J.R., Precise frequency power generation from an unregulated shaft, *Trans. A.I.E.E.* Pt. 2 *(Applic. and Ind.)* **79**, 442 (1960).
89. PLETTE, D.L. and CARLSON, H.G., Performance of a variable speed constant frequency electrical system, *I.E.E.E. Trans. on Aerospace* **AS-2**, 957 (1964).
90. VON ISSENDORFF, J., Der gesteuerte Umrichter, *Wiss. Veröff. Siemens-Werke* **14**, 1 (1935).
91. MEYER, M. and MÖLTGEN, G., Kreisströme bei Umkehrstromrichtern, *Siemens-Z.* **37**, 375 (1963).
92. VAN ECK, R.A., Frequency changer systems using the cycloconverter principle. *I.E.E.E. Trans. on Appl. and Ind.* **82**, 163 (1963).
93. LAWSON, L.J., The practical cycloconverter, *I.E.E.E. Trans. on Ind. and Gen. Appl.* **IGA-4**, 141 (1968).

94. Amato, C.J., An a.c. equivalent circuit for a cycloconverter, *I.E.E.E. Trans. on Ind. and Gen. Applic.* **IGA-2**, 358 (1966).
95. Amato, C.J., Sub-ripple distortion components in practical cycloconverters, *I.E.E.E. Trans. on Aerospace (Suppl.)* **AS-3**, 98 (1965).
96. Amato, C.J., Analog computer simulation of an s.c.r. as applied to a cycloconverter, *I.E.E.E. Trans. on Ind. and Gen. Applic.* **IGA-2**, 137 (1966).
97. Bland, R.J., Factors affecting the operation of a phase-controlled cycloconverter, *Proc. Instn elect. Engrs* **114**, 1908 (1967).
98. Langsdorf, A.S., *Theory of Alternating-current Machinery*, McGraw-Hill, New York, 1955.
99. Largiader, H., Design aspects of induction motors for traction applications with supply through static frequency changers, *Brown Boveri Rev.* **57**, 152 (1970).
100. Klingshirn, E.A. and Jordan, H.E., Polyphase induction motor performance and losses on nonsinusoidal voltage sources, *I.E.E.E. Trans. on Power App. and Systems* **PAS-87**, 624 (1968).
101. Chalmers, B.J. and Sarkar, B.R., Induction-motor losses due to nonsinusoidal supply waveforms, *Proc. Instn elect. Engrs* **115**, 1777 (1968).
102. Heumann, K. and Jordan, K.-G., Einfluß von Spannungs- und Stromoberschwingungen auf den Betrieb von Asynchronmaschinen, *A.E.G. Mitt.* **54**, 117 (1964).
103. Zweygbergk, S. and Sokolov, E., Verlustermittlung im stromrichtergespeisten Asynchronmotor, *Elektrotech. Z. Ausg. A* **90**, 612 (1969).
104. Jain, G.C., The effect of voltage waveshape on the performance of a 3-phase induction motor, *I.E.E.E. Trans. on Power App. and Systems* **83**, 561 (1964).
105. Robertson, S.D.T. and Hebbar, K.M., Torque pulsations in induction motors with inverter drives, *I.E.E.E. Petroleum and Chemical Ind. Conf., Los Angeles* (1969).
106. Ward, E.E. and Härer, H., Preliminary investigation of an inverter-fed 5-phase induction motor, *Proc. Instn elect. Engrs* **116**, 980 (1969).
107. McLean, G.W., Nix, G.F. and Alwash, S.R., Performance and design of induction motors with square-wave excitation, *Proc. Instn elect. Engrs* **116**, 1405 (1969).
108. Lipo, T.A., Krause, P.C. and Jordan, H.E., Harmonic torque and speed pulsations in a rectifier-inverter induction motor drive, *I.E.E.E. Trans. on Power App. and Systems* **PAS-88**, 579 (1969).
109. Lipo, T.A. and Krause, P.C., Stability analysis of a reluctance-synchronous machine, *I.E.E.E. Trans. on Power App. and Systems* **PAS-86**, 825 (1967).
110. Lawrenson, P.J. and Bowes, S.R., Stability of reluctance machines, *Proc. Instn elect .Engrs* **118**, 356 (1971).
111. Lipo, T.A. and Krause, P.C., Stability analysis for variable frequency operation of synchronous machines, *I.E.E.E. Trans. on Power App. and Systems* **PAS-87**, 227 (1968).
112. Nelson, R.H., Lipo, T.A. and Krause, P.C., Stability analysis of a symmetrical induction machine, *I.E.E.E. Trans. on Power App. and Systems* **PAS-88**, 1710 (1969).
113. Lipo, T.A. and Krause, P.C., Stability analysis of a rectifier-inverter induction motor drive, *I.E.E.E. Trans. on Power App. and Systems* **PAS-88**, 55 (1969).
114. Krause, P.C., Methods of stabilizing a reluctance-synchronous machine, *I.E.E.E. Trans. on Power App. and Systems* **PAS-87**, 641 (1968).
115. Fallside, F. and Wortley, A.T., Steady-state oscillation and stabilization of variable-frequency inverter-fed induction-motor drives, *Proc. Instn elect. Engrs* **116**, 991 (1969).
116. Krause, P.C. and Woloszyk, L.T., Comparison of computer and test results of a static a.c. drive system, *I.E.E.E. Trans. on Ind. and Gen. Applic.* **IGA-4**, 583 (1968).
117. Krause, P.C. and Lipo, T.A., Analysis and simplified representations of a rectifier-inverter induction motor drive, *I.E.E.E. Trans. on Power App. and Systems* **PAS-88**, 588 (1969).
118. Krause, P.C. and Lipo, T.A., Analysis and simplified representations of rectifier-inverter reluctance-synchronous motor drives, *I.E.E.E. Trans. on Power App. and Systems* **PAS-88**, 962 (1969).
119. Stepina, J., Betriebsverhalten der vom Wechselrichter gespeisten Asynchronmaschine, *Elektrotechnik MaschBau* **83**, 295 (1966).
120. Seefried, E., Ströme und Drehmomente eines umrichtergespeisten, flußgeregelten Drehstromasynchronmotors im stationären Betrieb, *Elektric* **20**, 301 (1966).
121. Sabbagh, E.M. and Shewan, W., Characteristics of an adjustable speed polyphase induction machine, *I.E.E.E. Trans. on Power App. and Systems* **PAS-87**, 613 (1968).
122. Ward, E.E., Kazi, A. and Farkas, R., Time-domain analysis of the inverter-fed induction motor, *Proc. Instn elect. Engrs* **114**, 361 (1967).
123. Li, K.Y.G., Analysis and operation of an inverter-fed variable-speed induction motor, *Proc. Instn elect. Engrs* **116**, 1571 (1969).

124. KRAUSE, P. C., Method of multiple reference frames applied to the analysis of symmetrical induction machinery, *I.E.E.E. Trans. on Power App. and Systems* **PAS-87**, 218 (1968).
125. KRAUSE, P. C. and HAKE, J. R., Method of multiple reference frames applied to the analysis of a rectifier-inverter induction motor drive, *I.E.E.E. Trans. on Power App. and Systems* **PAS-88**, 1635 (1969).
126. HEUMANN, K. and JORDAN, K.-G., Das Verhalten des Käfigläufermotors bei veränderlicher Speisefrequenz und Stromregelung, *A.E.G. Mitt.* **54**, 107 (1964).
127. NÜRNBERG, W., *Die Asynchronmaschine*, Springer-Verlag, Berlin, 1952.
128. RODEWALD, E., Betriebseigenschaften des Asynchronmotors bei Speisung mit niedrigen Frequenzen, *Siemens-Z.* **40**, Beiheft *Motoren für industrielle Antriebe*, 131 (1966).
129. DIXON, F. F. and TILEY, G. L., Low-frequency performance of a wound-rotor induction motor for mine hoist drives, *Trans. A.I.E.E.*, Pt. 3 *(Power App. and Systems)* **76**, 1140 (1957).
130. FISCHER, H.-J., Beitrag zum Betriebsverhalten von Asynchronmotoren bei veränderlicher Primärfrequenz, *Elektrie* **20**, 205 (1966).
131. O'NEIL, W. K., 20 kVA static frequency changer for direct drive induction servo motor, Instn elect. Engrs Conf. Publ. No. 37, *Conference on Servocomponents* 226 (1967).
132. HUGHES, F. M. and ALDRED, A. S., Transient characteristics and simulation of induction motors, *Proc. Instn elect. Engrs* **111**, 2041 (1964).
133. KRAUSE, P. C. and THOMAS, C. H., Simulation of symmetrical induction machinery, *I.E.E.E. Trans. on Power App. and Systems* **PAS-84**, 1038 (1965).
134. JORDAN, H. E., Analysis of induction machines in dynamic systems, *I.E.E.E. Trans. on Power App. and Systems* **PAS-84**, 1080 (1965).
135. JORDAN, H. E., Digital computer analysis of induction machines in dynamic systems, *I.E.E.E. Trans. on Power App. and Systems* **PAS-86**, 722 (1967).
136. ROBERTSON, S. D T and HEBBAR, K. M., A digital model for three-phase induction machines, *I.E.E.E. Trans. on Power App. and Systems* **PAS-88**, 1624 (1969).
137. DE SARKAR, A. K. and BERG, G. J., Digital simulation of three-phase induction motors, *I.E.E.E. Trans. on Power App. and Systems* **PAS-89**, 1031 (1970).
138. LAWRENSON, P. J. and STEPHENSON, J. M., Note on induction-machine performance with a variable-frequency supply, *Proc. Instn elect. Engrs* **113**, 1617 (1966).
139. ROGERS, G. J., Linearised analysis of induction-motor transients, *Proc. Instn elect. Engrs* **112**, 1917 (1965).
140. PFAFF, G., Zur Dynamik des Asynchronmotors bei Drehzahlsteuerung mittels veränderlicher Speisefrequenz, *Elektrotech. Z. Ausg. A* **85**, 719 (1964).
141. DE CARLI, A., MURGO, M. and RUBERTI, A., Speed control of induction motors by frequency variation, Paper 4C, *I.F.A.C. Congress, London* (1966).
142. WEST, J. C., JAYAWANT, B. V. and WILLIAMS, G., Analysis of dynamic performance of induction motors in control systems, *Proc. Instn elect. Engrs* **111**, 1468 (1964).
143. JAYAWANT, B. V. and BATESON, K. N., Dynamic performance of induction motors in control systems, *Proc. Instn elect. Engrs* **115**, 1865 (1968).
144. TSIVITSE, P. J. and KLINGSHIRN, E. A., Optimum voltage and frequency for polyphase induction motors operating with variable frequency power supplies, *I.E.E.E. Ind. and Gen. Applic. Group Fifth Annual Meeting* 815 (1970).
145. AMATO, C. J., Variable speed with controlled slip induction motor, *I.E.E.E. Industrial Static Power Conversion Conf.* 181 (1965).
146. MOKRYTZKI, B., The controlled slip static inverter drive, *I.E.E.E. Trans. on Ind. and Gen. Applic.* **IGA-4**, 312 (1968).
147. SALIHI, J. T., Simulation of controlled slip, variable speed induction motor drive systems, *I.E.E.E. Trans. on Ind. and Gen. Applic.* **IGA-5**, 149 (1969).
148. ABRAHAM, L. and KOPPELMANN, F., Käfigläufermotoren mit hoher Drehzahldynamik, *A.E.G. Mitt.* **55**, 118 (1965).
149. HUMPHREY, A. J., Constant horsepower operation of induction motors, *I.E.E.E. Trans. on Ind. and Gen. Applic.* **IGA-5**, 552 (1969).
150. AGARWAL, P. D., The GM high performance induction motor drive system, *I.E.E.E. Trans. on Power App. and Systems* **PAS-88**, 86 (1969).
151. HELMICK, C. G., How to specify adjustable-frequency drives (ii) selecting the inverter, *Control Engng* **14**, 83 (May 1967).
152. BECK, C.D. and CHANDLER, E.F., Motor drive inverter ratings, how to specify, how to analyze, how to compare, *I.E.E.E. Trans. on Ind. and Gen. Applic.* **IGA-4**, 589 (1968).
153. HELMICK, C. G., How to specify adjustable-frequency drives, (i) selecting the motor, *Control Engng* **14**, 69 (April 1967).

154. Barnes, E.C., Performance and characteristics of induction motors for solid state variable frequency drives, *I.E.E.E. Ind. and Gen. Applic. Group Fourth Annual Meeting* 653 (1969).

155. Chalmers, B.J., Design of induction motors supplied by thyristors, *Electl Rev., London* **184**, 525 (1969).

156. Gutzwiller, F.W., The current-limiting fuse as fault protection for semiconductor rectifiers, *Trans. A.I.E.E.*, Pt. 1 *(Communication and Electronics)* **77**, 751 (1958).

157. Jacobs, P.C., Fuse application to solid-state motor and power control, *I.E.E.E. Trans. on Ind. and Gen. Applic.* **IGA-2**, 281 (1966).

158. Helmick, C.G., Motor fusing for adjustable-frequency inverter systems, *I.E.E.E. Trans. on Ind. and Gen. Applic.* **IGA-5**, 40 (1969).

159. Dortort, I.K., Current-balancing reactors for semiconductor rectifiers, *Trans. A.I.E.E.*, Pt. 1 *(Communications and Electronics)* **77**, 452 (1958).

160. The Electricity Council, London, Harmonic distortion and permissible pulse number of consumers' rectifiers and inverters, Engineering Recommendation G. 5/2 (1967).

161. Ludbrook, A., Basic economics of thyristor adjustable speed drive systems, *I.E.E.E. Ind. and Gen. Applic. Group Fourth Annual Meeting* 591 (1969).

162. Wilson, T.G. and Trickey, P.H., D.C. machine with solid-state commutation, *Electl Engng, New York* **81**, 879 (1962).

163. Davis, R.M. and Jones, D.A., Conventional or electronic commutation?, Instn elect. Engrs Conf. Publ. No. 11, *Commutation in Rotating Machines* 93 (1964).

164. Krug, H., Der Ersatz des Kommutators der Gleichstrommaschine durch einen Stromrichter, *Elektrie* **22**, 104 (1968).

165. Brailsford, H.D., Solid state commutation for direct current motors, Paper No. 7-4, *I.E.E.E. Intermag. Conf.* (1964).

166. Moczala, H., Gleichstrom-Kleinstmotoren mit elektronischer Kommutierung, *A.E.G. Mitt.* **55**, 114 (1965).

167. Yates, W.W., Brushless d.c. motors, *Mach. Des.* **38**, 136 (1966).

168. Miyairi, S. and Tsunehiro, Y., Analysis and the characteristics of a commutatorless motor as a d.c. motor, *Electl Engng in Japan* **85** (No. 9), 51 (1965).

169. Miyairi, S. and Tsunehiro, Y., Damper winding in a commutatorless d.c. motor, *Electl Engng in Japan*, **87** (no. 8), 67 (1967).

170. Kübler, E., Der Stromrichtermotor, *Elektrotech. Z. Ausg. A* **79**, 15 (1958).

171. Ohno, E., Kishimoto, T. and Akamatsu, M., The thyristor commutatorless motor, *I.E.E.E. Trans. on Magnetics* **MAG-3**, 236 (1967).

172. Tsuchiya, T., Basic characteristics of cycloconverter-type commutatorless motors, *I.E.E.E. Trans. on Ind. and Gen. Applic.* **IGA-6**, 349 (1970).

173. Schmitz, N.L. and Long, W.F., The cycloconverter driven doubly-fed induction motor, *I.E.E.E. Conf.* Paper No. 70, TP 649-PWR (1970).

174. Haböck, A. and Köllensperger, D., Anwendung und Weiterentwicklung des Stromrichtermotors, *Siemens-Z.* **45**, 180 (1971).

175. Lamb, C.St.J., Commutatorless alternating-voltage-fed variable-speed motor, *Proc. Instn elect. Engrs* **110**, 2221 (1963).

176. Tustin, A., Electric motors in which commutation is by switching devices such as controlled silicon rectifiers, *Proc. Instn elect. Engrs* **111**, 151 (1964).

177. Bhagwat, P.G. and Wolfendale, E., Silicon controlled rectifiers replace commutator in d.c. motor, *Control* **8**, 24 (1964).

178. Wolfendale, E., A d.c. motor with s.c.r. commutator and synchronous speed control, *Control* **9**, 497 (1965).

179. Bhagwat, P.G., Speed control of d.c. shunt motor using thyristor bridge, *Control* **11**, 222 (1967).

180. Lamb, C.St.J., Dunworth, A. and Walker, B.S., Digital switching for speed control of commutatorless a.c. motor, *Control* **11**, 217 (1967).

181. Lamb, C.St.J., Analysis and testing of a direct-voltage induced-e.m.f.-commutated thyristor motor, *Proc. Instn elect. Engrs* **117**, 1975 (1970).

182. Zagalsky, N. and Shepherd, W., Wide-range reversible voltage controllers for polyphase induction motors, *Trans. A.I.E.E.* Pt. 2 *(Applic. and Ind.)* **81**, 272 (1962).

183. Bailey, F.M. and Konrad, C.E., Dynamic characteristics of controlled rectifier alternating current drives, Instn. elect. Engrs Conf. Publ. No. 11, *Commutation in Rotating Machines* 123 (1964).

184. Shepherd, W. and Stanway, J., The polyphase induction motor controlled by firing angle adjustment of silicon controlled rectifiers, *I.E.E.E. Internat Conv. Rec.* Pt. 4, **12**, 135 (1964).

185. KORB, F., Einstellung der Drehzahl von Induktionsmotoren durch antiparallele Ventile auf der Netzseite, *Elektrotech. Z. Ausg. A* **86**, 275 (1965).
186. PAICE, D.A., Induction motor speed control by stator voltage control, *I.E.E.E. Trans. on Power App. and Systems* **PAS-87**, 585 (1968).
187. TAKEUCHI, T.J., *Theory of SCR (Thyristor) Circuit and Application to Motor Control*, Tokyo Electrical Engineering College Press, 1968.
188. GUTT, H.-J., Bemessung stellergespeister Asynchronmaschinen für bürstenlose drehzahlsteuerbare Antriebe, *Siemens-Z.* **44**, 531 (1970).
189. FINZI, L.A. and PAICE, D.A., Analysis of the solid-iron rotor induction motor for solid-state speed controls, *I.E.E.E. Trans. on Power App. and Systems* **PAS-87**, 590 (1968).
190. ALGER, P.L., LUNEBURG, S.E. and MESTER, R.L., Alnico bar capacitor motor with s.c.r. speed control, Instn elect. Engrs Conf. Publ. No. 11, *Commutation in Rotating Machines* 128 (1964).
191. ALGER, P.L., COELHO, W.A. and PATEL, M.R., Speed control of wound rotor motors with s.c.r.s and saturistors, *I.E.E.E. Trans. on Ind. and Gen. Applic.* **IGA-4**, 477 (1968).
192. RAMA RAO, N. and JYOTHI RAO, P.A.D., Rerating factors of polyphase induction motors under unbalanced line voltage conditions, *I.E.E.E. Trans. on Power App. and Systems* **PAS-87**, 240 (1968).
193. KRAUSE, P.C., A constant frequency induction motor speed control, *Proc. National Electronics Conf.* **20**, 361 (1964).
194. KOVACS, K.P., Über die genaue und vollständige Simulation des am Ständer mit steuerbaren Siliziumtrioden geregelten Drehstrom-Asynchronmotors, *Acta Tech. Acad. Scient. Hungar.* **48**, 445 (1964).
195. NOVOTNY, D.W. and FATH, A.F., The analysis of induction machines controlled by series connected semiconductor switches, *I.E.E.E. Trans. on Power App. and Systems* **PAS-87**, 597 (1968).
196. SHEPHERD, W., On the analysis of the three-phase induction motor with voltage control by thyristor switching, *I.E.E.E. Trans. on Ind. and Gen. Applic.* **IGA-4**, 304 (1968).
197. LIPO, T.A., The analysis of induction motors with voltage control by symmetrically triggered thyristors, *I.E.E.E. Conf.* Paper No. 70 TP 661-PWR (1970).
198. BEDFORD, R.E. and NENE, V.D., Voltage control of the three-phase induction motor by thyristor switching: a time-domain analysis using the α–β–0 transformation, *I.E.E.E. Trans. on Ind. and Gen. Applic.* **IGA-6**, 553 (1970).
199. PETERSEN, J.G., Thyristor drives for a.c. cranes and hoists, *A.S.M.E.–I.E.E.E. Materials Handling Conf.* (1965).
200. SHEPHERD, W. and STANWAY, J., An experimental closed-loop variable speed drive incorporating a thyristor driven induction motor, *I.E.E.E. Trans. on Ind. and Gen. Applic.* **IGA-3**, 559 (1967).
201. LINGARD, B.W., JOHNSON, R.W. and SHEPHERD, W., Analysis of thyristor-controlled single-phase loads with integral-cycle triggering, *Proc. Instn elect. Engrs* **117**, 607 (1970).
202. ERLICKI, M.S., BEN URI, J. and WALLACH, Y., Switching drive of induction motors, *Proc. Instn elect. Engrs* **110**, 1441 (1963).
203. MEYER, M., The single-range induction motor and static converter cascade system—an excellent drive for small speed control ranges, *Siemens Rev.* **29**, 175 (1962).
204. ANSCHÜTZ, H., Der Einsatz von Stromrichterkaskaden für industrielle Antriebe, *B.B.C. Nachr.* **47**, 13 (1965).
205. ELGER, H. and WEISS, M., A subsynchronous static converter cascade for variable-speed boiler feed pump drives, *Siemens Rev.* **35**, 405 (1968).
206. SHEPHERD, W. and STANWAY, J., Slip power recovery in an induction motor by the use of a thyristor inverter, *I.E.E.E. Trans. on Ind. and Gen. Applic.* **IGA-5**, 74 (1969).
207. BLAND, R.J., HANCOCK, N.N. and WHITEHEAD, R.W., Considerations concerning a modified Kramer system, *Proc. Instn elect. Engrs* **110**, 2228 (1963).
208. MEYER, M., Über die untersynchrone Stromrichterkaskade, *Elektrotech. Z. Ausg. A* **82**, 589 (1961).
209. MIKULASCHEK, F., Die Ortskurven der untersynchronen Stromrichterkaskade, *A.E.G. Mitt.* **52**, 210 (1962).
210. ALBRECHT, S. and GAHLEITNER, A., Bemessung des Drehstrom-Asynchronmotors in einer untersynchronen Stromrichterkaskade, *Siemens-Z.* **40**, Beiheft *Motoren für industrielle Antriebe* 139 (1966).
211. MILJANIC, P.N., The through-pass inverter and its application to the speed control of wound rotor induction machines, *I.E.E.E. Trans. on Power App. and Systems* **PAS-87**, 234 (1968).
212. STÖHR, M., Vergleich zwischen Stromrichtermotor und untersynchroner Stromrichterkaskade, *Elektrotechnik MaschBau* **57**, 581 (1939).
213. PAICE, D.A., Speed control of large induction motors by thyristor converters, *I.E.E.E. Trans. on Ind. and Gen. Applic.* **IGA-5**, 545 (1969).

214. Shepherd, W. and Khalil, A. Q., Capacitive compensation of thyristor-controlled slip-energy-recovery system, *Proc. Instn elect. Engrs* **117**, 948 (1970).
215. Lavi, A. and Polge, R. J., Induction motor speed control with static inverter in the rotor, *I.E.E.E. Trans. on Power App. and Systems* **PAS-85**, 76 (1966).
216. Stöhr, M., Stromrichterkaskade für Doppelzonenregelung, *Elektrotechnik MaschBau* **58**, 177 (1940).
217. Ohno, E. and Akamatsu, M., Secondary excitation of an induction motor using a self-controlled inverter (super-synchronous thyristor Scherbius system), *Electl Engng in Japan* **88** (no. 10), 76 (1968).
218. Kazuno, H., A wide-range speed control of an induction motor with static Scherbius and Krämer systems, *Electl Engng in Japan* **89** (no. 2), 10 (1969).
219. Abraham, L. and Patzschke, U., Pulstechnik für die Drehzahlsteuerung von Asynchronmotoren, *A.E.G. Mitt.* **54**, 133 (1964).
220. Golde, E., Asynchronmotor mit elektronischer Schlupfregelung, *A.E.G. Mitt.* **54**, 666 (1964).
221. Koppelmann, F. and Michel, M., Kontaktlose Steuerung der Drehzahl von Asynchronmotoren mit Hilfe antiparalleler Thyristoren, *A.E.G. Mitt.* **54**, 126 (1964).

Index